GEOGRAPHICAL DATA

sources, presentation, and analysis

Hugh Matthews
Ian Foster

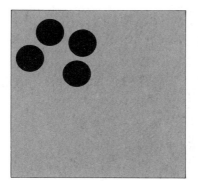

OXFORD UNIVERSITY PRESS

Oxford University Press, Walton Street, Oxford OX2 6DP

Oxford New York Toronto
Delhi Bombay Calcutta Madras Karachi
Petaling Jaya Singapore Hong Kong Tokyo
Nairobi Dar es Salaam Cape Town
Melbourne Auckland

and associated companies in
Berlin Ibadan

Oxford is a trade mark of Oxford University Press

© Oxford University Press 1989
Reprinted 1991
ISBN 0 19 913328 X

Set by Tradespools Ltd., Frome, Somerset
Printed in Great Britain by Cambridge University Press, Cambridge

ACKNOWLEDGEMENTS

The publishers and authors would like to thank the
following for their permission to use copyright
material:

Crown Copyright: p.37, 43,   David & Charles
(Publishers Ltd.): p.25,   Devon Record Office:
p.26,   OUP ©/Nick Fodgen: p.28,   Nigel Press
Associates: p.59,   Public Record Office, London:
p.30,   University of Dundee: p.45 (top and
bottom),   John Wiley & Sons Ltd./Journal of
Climatology: p.46.

Illustrations by Geoff Haddon.

Every effort has been made to trace and contact
copyright holders but this has not always been
possible. We apologise for any infringement of
copyright.

CONTENTS

Chapter 1 PREPARING A PROJECT

'Listen and know for a day, see and know for a week, do and know for a lifetime.'

1.1 GETTING GOING

Field study forms an important part of geography. In a field study you are undertaking your own research work. To do this effectively you must be able to:

  i show initiative and imagination in selecting a topic for inquiry;
  ii make decisions about geographically appropriate sources of information and strategies of data collection;
 iii classify, analyse, interpret and present data in a suitable form;
 iv draw conclusions and communicate findings in a clear manner.

Most field studies can be organised into a number of stages (Figure 1.1). This book sets out to examine some of these stages. It is intended to help the advanced level student make appropriate decisions when carrying out geographical investigations.

Figure 1.1
Stages in project design

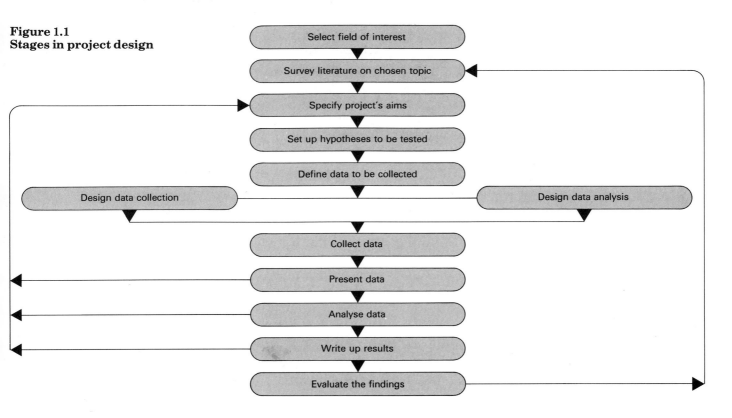

1.2 DEFINITION OF PRIMARY AND SECONDARY DATA

Geographical data come in many different forms. A basic distinction can be made between *primary* and *secondary* data.

Primary data refers to the sorts of information that can be collected first hand by fieldwork and questionnaire surveys.

1

Secondary data are those found in published sources, such as official statistics, maps and aerial photographs, or are gathered by some agency other than yourself.

Data are the raw materials of project work in geography. Many investigations will involve a combination of both primary and secondary data. It is important to make correct decisions about which sources of information are most appropriate. Attention should be given to such matters as availability, ease of collection and reliability. Some official data which are collected are *not* accessible to the public. For example, although the Census records details about individuals in a household, such information is protected by a law of confidentiality for one hundred years, and in the meantime only grouped data are available (the nature and form that these take is explained in Chapter 3). Sometimes no convenient secondary data source exists and so a survey has to be carried out. This can be time consuming and expensive, and careful thought needs to be given to setting up such an exercise. Whether dealing with primary or secondary sources consideration should be given to data reliability. Even the most official-looking statistics may contain errors or present details in a biased manner. Before placing faith in any data source try to find out how the data were collected as this will reveal some of their limitations.

1.3 WHY COLLECT DATA?

There are a number of important points relating to why we collect data in the first instance; it is important that you should think about what makes a good field study so that you collect data for the *right reasons*. Studies in human and physical geography should be based upon a sound 'scientific' approach to the problem – but what approach should we take? Is it enough to get the right sample or choose the most appropriate statistical technique for analysing the data we collect? These and other questions about the principles and methods of study should be thought about before we rush into the field to gather even more data or delve into historical archives for secondary sources of information.

Approaches to study

Two common approaches to geographical study are available to us. These are known as the *inductive* and *deductive* methods. It is important that you understand the differences between these two approaches and should try to decide which is the best to use for a particular type of study. They are summarised in Figure 1.2, and the following sections describe the major differences.

**Figure 1.2
The inductive and deductive methods of study**

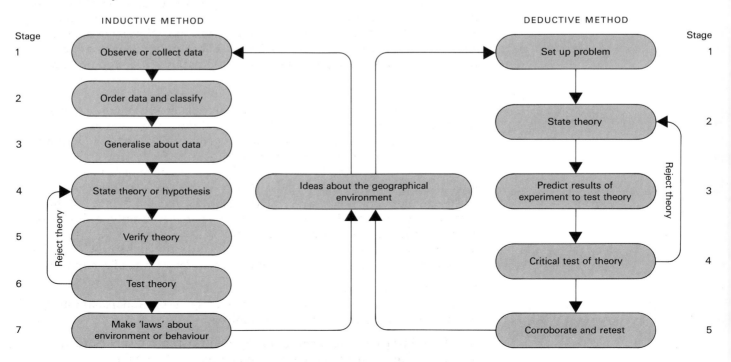

INDUCTIVE METHOD

Stage	
1	Observe or collect data
2	Order data and classify
3	Generalise about data
4	State theory or hypothesis
5	Verify theory
6	Test theory
7	Make 'laws' about environment or behaviour

Reject theory

Ideas about the geographical environment

DEDUCTIVE METHOD

	Stage
Set up problem	1
State theory	2
Predict results of experiment to test theory	3
Critical test of theory	4
Corroborate and retest	5

Reject theory

The inductive approach

The *inductive* approach, sometimes called the *classical* method, involves a number of stages summarised in the left hand column of Figure 1.2. The starting point is the observation of the environment or the collection of geographical data. Having collected the data, the second stage is to order and classify that data. We can see several examples of this in the field of biogeography and soils. Early studies of plants and animals led to the organisation of the natural world into the plant and animal kingdoms which were then subdivided into sub-kingdoms, classes, and orders. Soil scientists are also concerned with classification; for example in Britain we may find a number of soil types such as acid brown earths, brown earths and podzols. This classification is based on observed characteristics of horizons which develop in soil profiles. The third stage is to generalise about the data; the classification process in soil studies also includes this aspect because we will find a gradual progression from one soil type to another, yet we still need to place soils into particular groups.

On the basis of some collected data, we may find that certain relationships exist. For example, in a study of beaches we may see that the largest stones are found furthest from the water's edge. On the basis of this finding we can make a hypothesis (stage 4) which says that on stony beaches, the largest stones are always to be found at this point on the beach. Stages 5 and 6 in Figure 1.2 require us to carry out the same study on lots of beaches and, if we get the same results from these studies, we have established a law about how the environment behaves.

The deductive approach

The *deductive* approach, sometimes called the *critical rational* method, is an alternative to the inductive method and is shown in the right hand column of Figure 1.2. This method starts with the setting up of our initial problem. For example, we may wish to examine the same hypothesis as before about the size of stones on a beach. On the basis of what we know about coastal processes, we may set up a theory to test (stage 2). Before collecting the data, we should suggest what the outcomes of the experiment might be (stage 3) and we should then try to be as objective as possible by setting up our experiment in such a way that we try to prove our theory wrong (stage 4). If our theory is proved wrong, we turn it down and return to look again. If we do not reject the theory, it is possible to set up new experiments to test the same ideas in a different way. It is important to realise that just because a theory is not rejected it still need not be true. It may be possible that we designed a very poor experiment to test it in the first place and that, in a few years time, someone else may design a better means of testing the idea. This means that by using the deductive method, we do not obtain 'laws' about the way in which the world works but have hypotheses waiting to be tested in a 'rigorous' and 'scientific' way.

1.4 SAMPLING

With most project work it will be impossible to cover every member of a *population*. For example, a study of a National Park may require information on where people have come from and why they have chosen to visit this place. We cannot hope to obtain that sort of information for every visitor, even over the briefest time period. As a result some form of selection needs to be carried out and this is known as *sampling*.

The sample should be carefully chosen so as to be *representative* of the population from which it is drawn. It would be unwise simply to choose for interview the first fifty car drivers arriving at a car park within the National Park. In order to achieve a sample which is representative of the whole population the sample must be of correct *design* and of correct *size*.

Methods of sampling

Three principal sampling designs are commonly used: random sampling, systematic sampling and stratified sampling.

Random Sampling Random sampling enables a sample to be drawn from a population in a completely random way. Each member of a

TABLE 1.1
Extract from random number tables

20	17	42	28	23	17	59	66	38	61
74	49	04	49	03	04	10	33	53	70
94	70	49	31	38	67	23	42	29	65
22	15	78	15	69	84	32	52	32	54
93	29	12	12	27	30	30	55	91	87
02	10	51	55	92	52	44	86	42	25
11	54	48	63	94	60	94	49	57	38
40	88	78	71	37	18	48	64	06	57
15	12	54	02	01	37	38	37	12	93
50	57	58	51	49	36	12	53	96	40
45	04	77	97	36	14	99	45	52	95
44	91	99	49	89	39	94	60	48	49
16	23	91	02	19	96	47	59	89	65
04	50	65	04	65	65	82	42	70	51
32	70	17	72	03	61	66	26	24	71
69	85	03	83	51	87	85	56	22	37
06	77	64	72	59	26	08	51	25	57
27	84	30	92	63	37	26	24	23	66
55	04	61	47	88	83	99	34	82	37
22	77	88	33	17	78	08	92	73	49

RANDOM SAMPLING

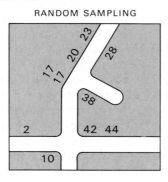

SYSTEMATIC SAMPLING

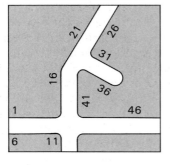

STRATIFIED SAMPLING

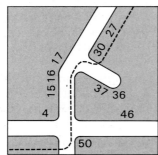

Figure 1.3
Sample design: 10 household strategy

population has an equal chance of being selected. To select from this population, each member is assigned a number. In order to ensure random selection it is necessary to use a random selection procedure such as random number tables (Table 1·1). The tables are produced by a computer, and are usually printed with numbers arranged in blocks of four for ease of reading. To choose the sample we can read off any series of numbers with any number of digits, although the number of digits used depends on the size of the population. For example, for a population of 100 you would use 3 digits, ignoring all those numbers of 101 and above and including all those of 100 and under in the sample; for a population of 1000 you would use 4 digits and so on. You can start anywhere within the table, reading the numbers either horizontally or vertically. From wherever a start is made it is important to continue reading the series or numbers next to each other. For example, a study is to be made of where housewives go to shop and 10 households on an estate are to be randomly selected for interview (Figure 1.3). There are 50 households in the population and so no more than two random number digits need be used. Starting at the top left hand corner on the random number tables (remember that you can start anywhere) a set of two digits is read off until 10 numbers fall in the range of 01 to 50: 20 17 42 28 23 17 (59) (66) 38 (61) 02 10 (51) (55) (92) (52) 44. Note that these numbers are produced by working progressively along a horizontal row. Households with these numbers now form the sample for interview.

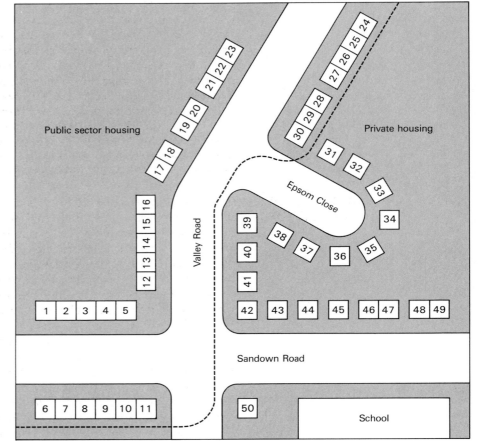

There are certain problems with this type of sample design. A sample of shoppers in a shopping centre could not be selected at random simply because the total population is not known and each shopper cannot be given a number. Equally in our household survey a random sample makes no allowance for different types of households known to be found on the estate. Also, the same member of a population can be selected more than once. The random sampling method is commonly used for selecting from populations in physical geography such as deriving information from maps or selecting materials in a field site.

Systematic sampling A systematic sample is one which is selec regular manner. Taking every fifth house on the estate would pr systematic sample (Figure 1.3). On the other hand, a sample surv land-use in an area may be drawn by taking a series of points loca the intersection of a 1 km grid (such as the Ordnance Survey grid).

Systematic sampling will provide an even cover of the population, in the sense that the sample members are selected at constant intervals and so it avoids the 'bunching' that may occur with random sampling. Although simple and easy to apply, this design may introduce *bias* into the results. Care must be taken to ensure that the regular series of sample points does not coincide with some other regular feature within the study area. Every fifth house may be an end of terrace or corner house and so may be untypical of the estate as a whole.

Stratified sampling Sometimes the population to be sampled is made up of different groups and the sample needs to take into account the relative proportion of these groups. For example, on the housing estate 20 of the households are owner-occupied, whereas 30 are in council tenancy. Stratified sampling will select a representative sample from these two groups and in a random way. In this instance if 10 households are to be selected, the correct balance is 4 owner occupiers and 6 council tenants (Figure 1.3). A random sampling design can be followed to select the members of each group.

The extent to which stratification can be carried out depends upon how much is known about the total population. This method is very useful for small samples, especially if a comparison is to be made of the different groups within a sampled population. Stratified sampling schemes are commonly used for collecting primary data about the physical environment. For example, a biogeographical study may try to find out whether differences occur in the number and composition of species in wet and dry heathland communities. Alternatively, a geomorphologist may want to test the idea that the size of pebbles varies according to the different levels of a beach. Here, some stratification is achieved on the basis of wet/dry habitat or high/low level beach. Within these subgroups a random or stratified sampling scheme may be used.

1.5 ERRORS OF DATA COLLECTION

Sampling will never be undertaken without some possibility of making errors. These errors occur for a number of reasons. If we try and count all of the pebbles in one section of a stream bed, for example, and repeat the exercise several times, it is unlikely that we will obtain the same answer on two occasions simply because of the difficulties of counting and remembering which ones have or have not been counted already. This is one form of error which is called *measurement error*. However, if two people counted the same set of pebbles and we again obtain a difference in the result, then the error this time will in part be a result of measurement error and in part a result of *operator error*. This often leads different people to measure the same thing with slightly different results. If one observer consistently over or underestimates the measurement, it is called *bias*. Both measurement and operator errors can be estimated if we are conducting a detailed investigation of a particular environment when one or more than one person is involved in the experiment. Here, it is possible to set up a controlled experiment to let all the people involved in sampling take the same measurements several times on a small subsample and find what differences result.

If we return to our stream bed sampling example, we may have decided to take 50 samples from the stream bed in order to calculate the average size of the sediment in the expectation that this sample will represent the population as a whole (all the pebbles in the section of river of interest to us). If we then select another 50 samples from the same site, and calculate the same values from it, we may again find small differences in the result. This is because we have a third source of error known as *sampling error*. Unlike measurement and operator error which can be found by setting up the right kind of experiment, sampling error can not be found by the same

5

means. However, the way in which statistical analysis handles the collected data (Chapter 5) allows us to work out the *probability* of that sampling error. This introduces some uncertainty into studies of all social, economic and physical phenomena, but we are still in firm control of the experiment because we choose the probabilities or odds which we are prepared to accept *before undertaking the experiment*. Of course, it may not be possible to take samples from the whole population because it is not available to us for some reason. On some occasions, we may specifically exclude some parts of the population for a given reason: for example we may only be able to measure particles of sediment in the bed of a stream which are larger than 2mm diameter with the technique we have available. The conclusions we come to are therefore not about the entire population but about a *purposive sample* of that population (Figure 1.4). When dealing with particle shape, we might also decide to exclude from our sample any pebbles which show evidence of mechanical fracture rather than stream abrasion. This will of course lead to an incorrect interpretation if we try to draw some conclusions about all of the particles on the stream bed.

The choice of a sampling procedure is therefore dependent upon the population we wish to sample. The population we want to sample is called the *target population*, but it is only possible to draw conclusions about the entire population if it is available for measurement. If not, we can only draw conclusions about the *sampled population* (Figure 1.4).

**Figure 1.4
Approaches to sampling in geography**

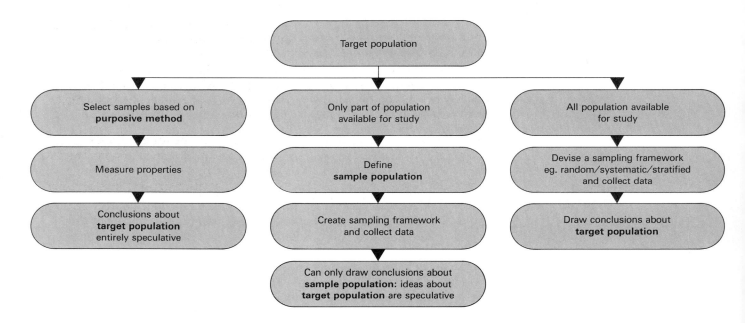

1.6
SAMPLE SIZE

There are statistical methods designed to calculate sample sizes in order to allow accurate statements to be made about the population from which the sample is drawn. In practice, for most studies, you will often have to limit your sample to what you can manage within a given time period. What this amounts to depends on the nature of the survey or the length of questionnaire, how many members make up the total population, and what form of analysis is to be carried out. As a rule of thumb, if conducting interviews or collecting data, aim for a sample of about 30 cases. Generalising about very small samples of 10 or less can lead to untypical and unrepresentative results. Very large samples create too much data to handle in a convenient way.

When planning the size of a sample for a questionnaire you must expect *non-response*. There is often a considerable difference between a *selected* and *achieved* sample. Even the most experienced interviewers would expect no more than an 80–90 per cent response rate. For inexperienced interviewers the success rate may fall to 50 per cent. Try to think ahead along these lines and select a sample of at least *double* the required number.

At the end of the day it is very hard to predict the likely response rate. If conducting a house survey ensure that you keep a note of where you have visited. It may be necessary to set up a second or 'booster' sample to complete the survey.

1.7
LEVELS OF
MEASUREMENT

The exact nature of data collected depends upon the *level* at which measurement is made. This is important because it will often determine the type of statistical test we might want to use in order to analyse the data. There are four levels of measurement:

The **Nominal** scale is a scale based upon classification of data into mutually exclusive groups or categories. No attempt is made to *quantify* the size of the groups. Objects are classified on the basis of *one* of the identified groups. Examples include classification of different rock types, land use, and ethnic groups.

The **Ordinal** scale measurement allows us to rank the classes, but *not* to distinguish between them on the basis of the size of a measured variable. For example, we may classify soils on the basis of a colour scale from light to dark. In geology, Mohs' scale of hardness arranges specimens on an ascending scale with unequal intervals, and in perception studies you may wish to establish locations where people 'prefer' to live or shop.

The **Interval** scale of measurement is used when we have equal intervals between the measurement units but without regard to an absolute zero. The most obvious example is the Celsius and Fahrenheit temperature scales. Here, 40°C is not 'twice as hot' as 20°C; these temperatures are the same as 104° and 68° on the Fahrenheit scale. Also, the relationship between the two scales requires a relatively complicated calculation; to convert °C to °F, we multiply by 9/5 and add 32.

The **Ratio** scale of measurement is different from the interval scale because it has an absolute zero, such as in measurements of length or area. Here, 20cm is twice as long as 10cm and measurements in one unit (e.g. inches) can be converted into other units (e.g. centimetres) by multiplying by a constant (in this case 2.54).

Chapter **2** PRIMARY DATA SOURCES

Primary data sources in human geography

2.1
TYPES OF SURVEY

There are two main ways of collecting your own data:

i *to observe what is going on*. This includes pedestrian and traffic counts, assessments of landscape quality, the mapping of different types of land use, and compiling registers of local services. These surveys involve no interviews and only gather information on parts of the built and natural environment that can be seen.

ii *to undertake questionnaire surveys*. These may be concerned with who people are, how they live and what they think. Some surveys are *descriptive*, noting such characteristics as where people go to work and by what form of transport. Others are *explanatory* and try to explain what is going on or why people behave in a certain way. In practice many surveys have both descriptive and explanatory parts.

2.2
QUESTIONNAIRE DESIGN

Purpose and content

Questionnaire surveys are the best method of obtaining information, usually in a statistical form, from a sample population in a relatively short time. Before deciding upon a survey, make sure that a questionnaire survey is worth planning. At this early stage do not be over-ambitious in what you set out to achieve. Avoid the temptation of covering everything that seems interesting. Each question should have a particular purpose. There is no room for questions which *may* come in useful. This means that you must be very clear about what issues you are going out to investigate. A useful practice is to list the questions against each item of interest. This will provide a check upon whether each issue is being covered adequately and help to prevent irrelevance.

Exploratory interviews

Before designing the questionnaire it is often worthwhile carrying out a few exploratory interviews in the study area. This will allow you to see whether the sorts of questions you are intending to ask cover the issues or whether additional sections need to be added.

Identifying the population

An important part of survey work is identifying the population to be interviewed. Sometimes this is straightforward. If you are interested in what sports local schoolchildren aged 11 to 18 participate in, then that is the population from which the sample is taken. On other occasions the population may be clearly defined, for example, senior citizens living alone, but the problem is how can this population be identified and reached? In this instance contact with the local Social Services Department may help you trace some of this group, but the people that they would be aware of form only a small part of the total population; they may only be able to identify old people receiving 'meals on wheels'. Often these difficulties cannot be overcome and your sample will be drawn from an incomplete population. What is important is that you appreciate the limitations of the survey and that you quote the results correctly. For example, the survey may not reflect the views of 'old people t[...]' 'local senior citizens receiving meals on wheels'.

Always attempt to define the population before setting out t[...] data. If conducting a house-to-house survey do not just intervie[...] person who answers the door. Be clear about whom you wish to i[...] Is it, say, the chief wage earners, or married women with childre[...]ged under 5 years? Those people who are asked to answer questions are usually referred to as *respondents*.

If conducting a random sample of adults the standard procedure is to use the Register of Electors, which provides the names and addresses of every person registered to vote, generally listed in street order. These lists are found in large local libraries and are also available from the local authority.

Types of data

Most questionnaires will probably include three different types of data:

i   *background* or *classificatory data* – this enables you to describe the sample in terms of such things as their age, sex, occupation, length of residence;

ii *activity data* – this informs about what people do, how often and when;

iii *attitudinal data* – this relates to what people think about an issue and gives them a chance to air their views.

Types of question

There are two types of questions, *open* and *closed*. Open questions are those which allow respondents to answer freely in their own words, such as, 'How satisfied are you with your standard of living?' Closed questions are those where the possible alternative answers are fixed in advance: for example, 'How long have you lived in this village: i) <1 year, ii) 1–5 years, iii) 6–10 years, iv) >10 years?' Both sorts of questions have their advantages and problems. Open questions are good in that people can say or write what they think. But they can take a long time to answer and replies vary so widely that analysis may be difficult. Also, it may be hard to take down exactly what people are saying. Closed questions are quicker and easier to process, especially if each category is numbered to permit quick running totals. But care must be taken to select a sufficiently wide set of alternatives from which people can choose. If options are too few people may be forced into answers they do not really mean and if options are too many then confusion may result.

When deciding upon the type of questions, consider the form of the analysis. For example, with closed questions it is possible to total how many people gave a particular answer. These totals are convenient for such tests as chi-squared analysis, which enables you to see whether members of different populations respond in the same or different ways to each other (see Chapter 5).

Wording and phrasing

In all questionnaires great trouble must be taken in wording the questions, as they need to be clear, simple and straightforward to understand.

Clear questions Always avoid jargon and assume nothing. The respondents answering the questions are unlikely to know a great deal about geography, so common geographical terms will mean nothing. For example, if carrying out a shopping survey people cannot be expected to know about 'central places' or 'high and low order goods'.

Simple questions Ensure that questions can only be interpreted in one way. 'When did you leave school?' may be answered as 3.30 p.m. or 17 years depending on the age of the person answering the question. 'At what age did you finish full-time education?' may be a much better alternative, especially as many people go on to take further education courses. This phrasing still leaves out those studying on a part-time basis, but you cannot hope to cover everything in one question.

Phrase questions so that they do not lead respondents to give a particular answer. For example, 'You don't think..... do you?' is asking for a 'No'; whereas 'Should not something be done about....?' is likely to get a 'Yes'.

Straightforward questions Common pitfalls are *double-choice* and *double-barrelled* questions. Double-choice questions call for more than one decision at a time. For example, 'Would you say that there is more violent crime now than ten years ago?' is asking for two assessments, more or less, now and then. A better phrasing would be 'Is there more violent crime now or was it higher ten years ago?'

Double-barrelled questions are those which ask several questions at the same time. For example, 'Do you think that the Government should increase or decrease its spending on education and defence?' Since people may hold different views about education and defence the question should be split into two.

Setting questions

Background questions With some background questions, such as age or length of residence, it is useful to ask people to give their replies in terms of broad categories. For example, 'How old are you: 0–15, 16–35, 36–50, 51–59, 60–65, 65+?'

On the other hand, when asking people to give their 'occupation' attempt to get a precise answer. Occupational groups provide a very useful way of classifying the sample, but this can only be achieved if answers are sufficiently detailed. This may call for some probing on the part of the interviewer. For example, to such answers as 'civil servant' or 'I work at...', follow up with 'What kind of work do you do?' Ask the rank of anyone in the police or armed forces. If someone describes themselves as a 'manager' or 'director' find out how many people they employ. The description 'engineer' covers many trades; try to find out whether the person is qualified or not.

Some background questions are better not asked. For example, 'income' is a sensitive issue. People may resent such a question or simply provide inaccurate figures in an attempt to impress the interviewer.

Memory questions Questions which require a good memory are prone to error. It is always better to ask questions about short time periods rather than long ones. For example, if carrying out a shopping survey it would be unrealistic to ask 'What towns have you visited to shop in the last year?' If the time period was changed to '...in the last month?' more meaningful answers would result. If a longer time period is absolutely necessary people often find it easier to recall events if these are linked to some part of the year. Rather than ask 'In the last six months, how many times have you visited the doctor?' change the wording to 'Since Christmas...' or 'Since Easter...'.

Attitude questions In relation to attitude questions, where people are asked to indicate agreement or disagreement with a question, it is often useful to use a *graded scale* rather than just 'Yes' or 'No'. A graded scale can allow for more varied answers. For example, 'Do you think more money should be spent on policing the inner city: i) strongly agree ii) agree iii) not sure iv) disagree v) strongly disagree?' Other scales can be used. 'How satisfied are you with local bus services: i) very satisfied ii) satisfied iii) not sure iv) unsatisfied v) very unsatisfied?'

The chart below illustrates another form of scaling. The question here is 'How would you describe life in this village?'

	2	1	0	−1	−2	
Friendly						Unfriendly
Open to newcomers						Closed to newcomers
Good place to bring up children						Poor place to bring up children

This form of scaling uses pairs of opposites (*bi-polar adjectives*) combined with a point scale to record people's strength of feeling towards an issue. In our example, describing aspects of life in a village. In the chart, strength of feeling can be scored positively or negatively. If the 'pairs' have no relevance a zero can be awarded.

Length

Long questionnaires should be avoided. Remember that most people will be busy and will have little time to spare. Usually a questionnaire should consist of no more than 30 questions and take less than 15 minutes to answer. Anything above this length and you run the risk of impatience and increasingly flippant answers. By keeping the questionnaire brief and to the point you are more likely to keep people's interest and gain their help. Think about where and when you will be asking the questions and this will give an idea of whether or not you have set too many.

Layout

Considerable care should be given to the layout of the questionnaire. Avoid asking all the background questions at the beginning or lumping them together later on. People are often very sensitive about their age or occupation so get things going before you ask for such details.

In the first place gain people's interest. Start with some easy lead-in questions and build up to more complex issues, like opinions and attitudes. Background information should be introduced at different times. Perhaps point out that this sort of detail is useful so that you can group together answers from people with similar backgrounds.

Presentation is very important. Response rates are likely to be higher and answers more carefully considered if the questionnaire is neatly typed and clearly set out. If asking people to write down answers, especially to open questions, the amount of space that is provided on the answer sheet will influence their replies.

2.3
DATA COLLECTION

Pilot survey

Even with the most carefully planned questionnaire there are bound to be some problems which you have not expected. Ideally, any questionnaire should be carefully *piloted*. That is, the draft questions should be tried out on a small group of respondents, no more than five, drawn from the same population as the actual ones would be and any difficulties picked up and corrected. This should highlight any misleading or confusing questions and draw attention to any major flaws in design.

Survey methods

There are a number of different ways by which data can be collected. The first decision is whether questionnaires should be distributed by post or hand.

Postal surveys can reach a large number of people with relatively little effort but they are very expensive and response rates are low. Also there is no guarantee that they will be answered by the person at whom the questionnaire is aimed. For most student projects this type of survey is not feasible.

More likely is the *hand-distributed questionnaire*. Two methods are commonly used: the *drop and collect survey* where questionnaires are left with people to fill in and collected at a later date, and the *face to face interview* in which the interviewer stays with the respondent to complete the questionnaire.

Both of these procedures have strengths and weaknesses. If the questionnaire is short and simple to understand the former method provides a quick and convenient way to scan a good size sample. However, interviewers can get a higher response rate, can ask supplementary questions and can clarify issues if there is any misunderstanding. But they may also affect how respondents answer the questions, either by an *interviewer effect,* such as the desire to please the interviewer, or through the way respondents react to the manner of the interviewer.

Locating the interview

The setting in which an interview takes place is another important consideration. If conducting a shopping survey on a high street many people will be rushing by, often laden with goods. Choose a position where people can sit down or rest whilst answering the questionnaire, such as near a street bench.

Door-to-door interviews may seem intimidating, but they offer a very successful way of collecting primary data. Remember that people are less likely to be willing to stop on a doorstep on a chilly winter's evening.

Some interviews may need to be carried out on private property, for example, the car park of an out-of-town shopping centre or in a shopping arcade. If this is the case make sure that you gain permission to undertake the survey beforehand.

Timing the interview

The time and day on which an interview is conducted may influence who takes part in the survey and likely response rates. For example, if a survey of shoppers is carried out on the morning of a weekday the people turning up at the centre may be very different to those attending the same place on an evening or a Saturday morning. On the first occasion there may be a large number of women with young children, during the evening there may be more men and youths amongst the shoppers, whilst on a Saturday families may compose the largest proportion. Similarly, if carrying out a doorstep survey, who is likely to be home will differ according to the time of day.

You can, of course, interview at various times, but Sunday morning between 10.00 and 12.00 a.m. is a good time, and so is Saturday afternoon and most other evenings between 6.30 and 8.30 p.m. Do not call after 9.30 p.m. or before 9.00 a.m. Think about who you want to take part in the survey, and their likely whereabouts during the day, as this will influence when and where you carry out the interview.

Conducting the interview

Before starting the questionnaire tell people who you are and the purpose of the survey. A good idea is to send an initial letter explaining the background to the project and asking for cooperation when you call. If people want to know why they were chosen you can reply that you picked their address at random. Above all, stress the *confidential* nature of the inquiry. No names or addresses should be recorded. Point out that you are only interested in the total number of replies to a question and not the particular view of an individual.

At all times appear confident and well organised. Casual presentation will result in a lot of refusals and careless answers. Ask the questions in a friendly, uncritical way. Never show surprise at an answer or voice your agreement or disagreement. Look interested in people's answers and never try to rush a reply.

Recording the interview

In order to keep the interview flowing smoothly and for ease of analysis it is often convenient to take down answers on a pre-coded sheet (Figure 2.1). This works best when closed questions are asked. On this example a running total of replies can be easily provided.

Completing the interview

At the end of the interview check that you have covered all parts of the questionnaire and be sure to thank the respondents for their help. If permission has been granted by an agency, on completion of the survey it is a good idea to send a letter thanking them for their cooperation and informing them that the work has been successfully carried out.

2.4
SOCIAL CLASSIFICATION

Why classify?

A major part of survey analysis is to see whether variation exists amongst *groups* of people *within* a sample with respect to their behaviour or attitudes. A starting point is to decide whether you are interested in

Questions						

PRE-CODED SHEET

RESPONDENTS

Questions	Questions	1	2	3	4	5	Total
1 Do you use public transport?	**1**						
i Yes	i	✓	✓	✓	✓	✓	5
ii No	ii						
2 If yes, how frequently do you use it?	**2**						
i daily	i	✓		✓			2
ii >twice a week	ii		✓				1
iii twice a week	iii				✓		1
iv once a week	iv					✓	1
v once a fortnight	v						0
vi <once a fortnight	vi						0
vii other	vii						0
3 How satisfied are you with the bus service?	**3**						
i very satisfied	i			✓			1
ii satisfied	ii		✓				1
iii dissatisfied	iii	✓			✓	✓	3
iv very dissatisfied	iv						0
v don't know	v						0
4 Are there any problems with the bus service?	**4**						
i frequency	i	✓					1
ii cost	ii				✓		1
iii routes	iii		✓	✓			2
iv reliability	iv	✓					1
v other	v						0

Figure 2.1
An extract from a questionnaire showing a completed pre-coded sheet

groups of individuals or groups of households with similar characteristics. Commonly three sorts of classification are used to identify these groups: *individual classifications,* based on demographic, educational, and employment characteristics of respondents; *household classifications,* based on social class and accommodation characteristics of households; *geographical classifications,* based on regional and areal characteristics of respondents or households.

The choice of classification depends upon the purpose of the survey and what sort of background data have been collected. For example, it may be important to show how age and sex influence the types of recreational activity in which people take part (individual classifications), on the other hand you may wish to see how social class and place of residence influences where households go for their summer holidays (household and geographical classifications). Whatever method is used there are no rigid rules which can be followed; each classification is open to interpretation. Also, classifications can be modified to suit the needs of the project. You may just be interested in how many boys and girls under 12 years of age smoke cigarettes in relation to those over 12 years and under 18 years, and so the classification needs only to recognise those age groups and the sex of the respondents.

In the following section some common ways of grouping populations are outlined. Many of the categories suggested can be combined or adjusted to meet the specific requirements of your survey.

Individual classifications

These classifications are based on the characteristics of the respondents.

DEMOGRAPHIC CHARACTERISTICS

Sex: male and female.

Age: different categories can be put forward. It is often useful to start with a detailed classification which can be collapsed together for analyses.

With children it is usual to conform to school stages:
   <5 years, 5<12, 12<17.

Adults can be grouped together into different stages of their life cycle:
   17<25 years, 25<45, 45<65, 65+.

Each category needs to be self-contained with no overlap at the end points of each range.

Marital status: this is a fairly straightforward grouping:
   single, married, widowed, divorced.

EDUCATIONAL CHARACTERISTICS

The problem with educational classifications is that they have to be flexible enough to cover people educated under different systems, that is, at different times this century and in different regions of the United Kingdom. Two main sorts of data are usually recorded: age at which full-time education was completed and highest educational qualification.

Age at which full-time education was completed:

   $\leqslant 14$, 15–16, 17–19, 20–24, >24.

Highest educational attainment:

No qualifications	A Level
Local school leaving certificate	Degree or above
Matriculation	Professional
O Level/CSE	Vocational
GCSE/16+	Other

EMPLOYMENT CHARACTERISTICS

This classification is based upon whether or not people are in employment.

Activity status:

Working full-time (>30 hours per week)
Working part-time (<30 hours per week)
Not working (seeking work, sick, retired, student, housewife)

Standard Industrial Classification: the revised Standard Industrial Classification (SIC), published in 1968 by the Central Statistical Office, provides a means of classifying different types of employment by work place. The full range of activities is first divided into broad Divisions, each denoted by a single digit from 0 to 9:

Digits	Divisions of the SIC
0	Agriculture, forestry and fishing
1	Energy and water supply industries
2	Extraction of minerals and ones other than fuels: manufacture of metals, mineral products and chemicals
3	Metal goods, engineering and vehicle industries
4	Other manufacturing industries
5	Construction
6	Distribution, hotels and catering; repairs
7	Transport and communication
8	Banking, finance, insurance, business services and leasing
9	Other services

Each Division is sub-divided into Classes (two digits), the Classes into Groups (three digits) and the Groups into Activity headings (four digits). There are 10 Divisions, 60 Classes, 222 Groups and 334 Activity headings.

Household classifications

The Census defines a household as 'a group of people who live regularly at the same address and who are *catered* for by the same person'. If other people live at the same address and are catered for by someone else or cater for themselves then they form a separate household.

Many surveys are interested in comparing the attitudes, opinions and behaviour of households with different backgrounds. Usually a spokesperson is chosen from the household to answer the questionnaire. Who is chosen depends upon the *target population* of the survey.

SOCIAL CLASS

A common way to distinguish between households is to find out the *occupation of the chief wage earner*. Nearly all surveys use occupation as a measure of social class. This is because occupation is a good indicator of income, education, life style, interests and beliefs. In this way households can be categorised by their social class characteristics. It is usual to use the occupation of the chief wage earner as a means of assessment regardless of whom you interview in the household or how many other members of the household are working.

A number of different classifications use occupational groupings as a way of recognising social class.

Registrar General's Social Classification: this is perhaps the most commonly used classification. The Registrar General provides a list of 20 000 occupational titles, which are grouped into 200 occupational units. These in turn can be arranged into 6 broad categories:

I	PROFESSIONAL	Accountants, judges, doctors, clergy, lecturers
II	INTERMEDIATE	MPs, nurses, teachers, pilots, farmers
IIIN	SKILLED NON-MANUAL	Clerical workers, secretaries, sales people
IIIM	SKILLED MANUAL	Carpenters, bus drivers, bricklayers
IV	PARTLY SKILLED	Postmen, production line workers, agricultural labourers
V	UNSKILLED	Labourers, window cleaners

Managers and foremen are raised a class.

Social grade: some market research surveys use social grade as a measure of social class:

AB Managerial, administrative or professional occupation (at a senior or intermediate level)
CI Supervisory or clerical occupation (ie. white collar) and junior managerial, administrative or professional
C2 Skilled manual worker
DE Semi-skilled and unskilled manual workers and state pensioners, widows (with no other wage earner in the household) and casual workers

ACCOMMODATION

For some surveys it is useful to classify households according to the tenure and type of their accommodation.

Tenure: three principal tenure categories can be identified: households in *owner-occupation* (mortgagors, leaseholders, outright owners); households *renting from local authority* (council tenants); and households in *private*

15

renting (renters in furnished and unfurnished accommodation, housing association tenants, tenants whose accommodation is tied with employment).

Type: households can be classified according to the age and form of their dwelling:

	Detached	Semi-detached	Terraced	Flat	Rooms
Pre-1918	1	2	3	4	5
Inter-war	6	7	8	9	10
Post-war to 1980	11	12	13	14	15
Recent since 1980	16	17	18	19	20

Further differences can be added by taking into account tenure.

Geographical classifications

There are many ways of classifying householders or respondents in terms of 'where' they live. The *scale* at which comparison is to be made is important. Classifications can range from large regions to small areas within urban centres. Also, different *types* of region can be recognised, such as economic planning regions, Area Health Authority regions or utility regions of the Gas, Water and Electricity Boards. Often the nature of the survey determines the classification to be used.

Regional classifications: at a broad scale the *Standard Region* classification is useful (Figure 2.2).

Areal classification: on a smaller scale than the region, the area in which the respondents live can be classified at different levels: county, local authority, ward, enumeration district. Often you may have to devise your own way of classifying areas. For example, a survey may want to compare an inner city with a suburban population, or an area of new, privately-owned housing with an old local authority housing estate.

**Figure 2.2
Standard Regions of the
United Kingdom**

Primary data sources in physical geography

Primary data in physical geography can be collected in the field by either measuring at the site of interest or collecting samples which we return to the laboratory for further analysis. These data can also be collected from experiments which are undertaken in carefully controlled laboratory conditions. The different types of primary data available for study in physical geography are shown in Figure 2.3.

**Figure 2.3
Primary data sources in physical geography**

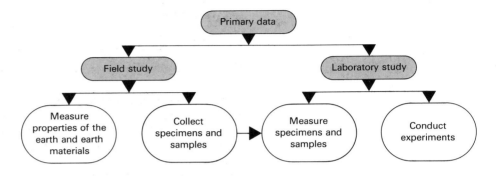

2.5 ACCURACY AND PRECISION

The data derived from field or laboratory study should be both *accurate* and *precise*. These terms mean different things but are often confused. By accuracy, we mean that the value or number obtained from the measurement of an object, say the diameter of a pebble, is very close to the 'true diameter'. By precision, we mean that if we measure the same object several times, the differences between each measurement are small. It is most desirable to have both high levels of accuracy and precision but we must recognise, especially if the method we use gives us a biased result, the possibility of having precise but inaccurate data. Similarly, the method we use may produce substantially different numbers if we repeat the measurement (imprecise) but the average of a large number of measurements could be close to the true value (accurate). Before collecting any data, make sure that you have found the best technique to use which gives precise and accurate information.

In physical geography we are concerned with many different types of data, and the following section discusses this problem in some detail.

2.6 MEASUREMENT: DIMENSIONS, UNITS, AND SCALES

It is only possible to look for order in the natural environment by careful measurement and collection of data. In order to do this it is important that we all use a standard system for measurement otherwise it is very difficult to compare our results with those published in books and scientific reports.

Dimensions

Geographers tend to gain most data from the measurement of three basic properties of the environment and earth materials. These are:

 i Distances and lengths (L)
 ii Units of time (T)
 iii Units of mass (M)

On the basis of these dimensions, most other variables can be calculated. For example, *area* is $L \times L$ or L^2 whereas *volume* is $L \times L \times L$ or L^3. Time is introduced as a dimension of measurement when we calculate velocity, discharge and acceleration. For example, velocity is $L \div T$, this is usually expressed as LT^{-1}, which is also the same as L/T. *Discharge* is a volumetric measurement per unit of time, $L^3 T^{-1}$, whereas *acceleration* is the change of velocity per unit of time in units of $L T^{-2}$. On some occasions, it is important that we consider *forces* operating in the natural environment such as those promoting sediment movement in a gravel-bed river or on an unstable slope. Force is also important because pressure (such as atmospheric pressure) is calculated from the application of a force

TABLE 2.1

Quantity	Name	Symbol
length	metre	m
mass	kilogram	kg
time	second	S
thermodynamic temperature*	Kelvin	K
electric current	ampere	A
luminous intensity	candela	cd
amount of substance	mole	mol

* note that water freezes at 273.15°K and boils at 373.15°K. This is a particularly awkward reference scale and part of it, the Celsius scale, is usually used. This scale sets the freezing point of water at 0°C and the boiling point at 100°C. The Kelvin Scale and Celsius scale therefore have the same number of units (100) between the freezing and boiling points of water.

2.7 EXPERIMENTAL CONTROL

per unit area. We cannot, of course, define force in terms of L and T alone, because it involves movement of an object, or body, of a known mass (M). This therefore introduces our third important dimension. Force is defined as MLT. Pressure is calculated as

$$\frac{\text{force}}{\text{area}} = \frac{MLT^{-2}}{L} = ML^{-1}T$$

Units and scales of measurement

Of course, we have not specified in the above discussion what units we are using in order to make our measurements of L, T and M. There are several units that could be chosen, for example:

Dimension	Units
L	Inches, feet, metres
T	Seconds, minutes, hours
M	Ounces, pounds, kilograms

If we mix up the units of measurement as shown above, it is difficult to convert one unit into another. There are several different reference systems which have been used in different countries throughout the world, but most recently the SI system has been adopted as the international reference system (SI stands for 'Le Systeme International d'Unites'). This system has seven so-called *base units*, from which all others are derived (Table 2.1).

Often with project work in science it is useful to introduce some *experimental control*. In the chemical laboratory, of course, we can repeat experiments under carefully controlled conditions to see whether we get the same results every time. Unless we carry out similar types of experiments, such as by using laboratory flumes to study river channels, it is unlikely that we can have such control over what we test in physical geography. This does not mean that the tests we do in the field are any less important, but it makes interpretation more difficult because another element of uncertainty is introduced. The best way to illustrate the use of a carefully controlled experiment is to look at an example of some of the methods available to us in order to assess the impact of human activity on the behaviour of river channels.

Case study: river channel change and urbanisation

From our own experiences we might suspect that the urban hydrological cycle will be different from the rural cycle, as shown in Figure 2.4. Also from our experience we might suggest that because the runoff from the urban area is higher, it is likely that the size of the river channels will also increase to carry a larger volume. Think of the different ways we could test the following hypothesis:

'The increase in the runoff from urban areas will cause river channels to increase their natural size downstream of the urban area.'

There are three methods we might use to test this hypothesis:

i *A before and after study:* This means that in a situation where we know that an urban development is about to occur, we can measure the river channels before and after building is completed. This type of study is difficult because we may have to wait some months or even years before we can repeat the measurements and complete the study.

ii *A paired catchment experiment:* This means that we choose two drainage basins, preferably of similar size, geology, relief and vegetation, and the non-urbanised catchment acts as our experimental control. This method can be used readily without the need to wait for any development to be started or completed. Any differences between the rivers in the two basins might therefore result from the effects of urban runoff.

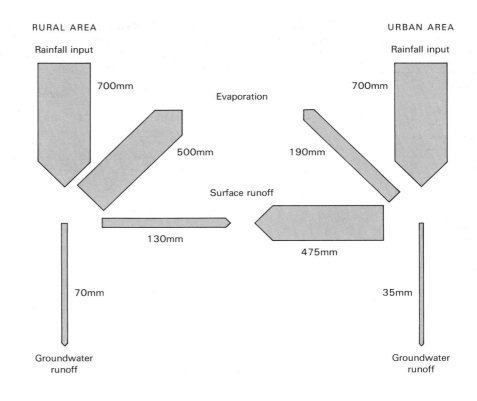

Figure 2.4
The water balance of the urban and rural areas of Moscow

iii *A single catchment experiment:* Where the previous two methods
cannot be used, we may only have access to the catchment which has
already been urbanised. We could measure the size of channels above
and below the urban area but of course direct comparison is impossible
because as we move downstream the drainage basin area increases,
which also makes channel size increase. Here we must make use of a
less direct approach. For several points in the river channel upstream
of the urban area it is possible to measure both channel cross sectional
area and drainage basin area. Plot one against the other on a graph as
shown in Figure 2.5. Draw a best fit line by eye through the data.
Record the same information downstream of the urban area and plot
this on the same graph. Project the best fit line and see if the
downstream data lie close to it. If not, we might suspect that the
channels downstream are 'larger than expected'. This method shows
the use of experimental control to remove the effects of another
variable (in this case drainage basin size) which also influences the size
of the river channel.

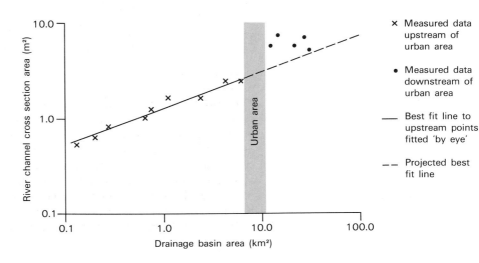

Figure 2.5
The effect of urban areas on river channel size

Carrying out the study

Having set out the initial hypothesis, we might choose method three as the framework for collecting the necessary information. The next task is to think about scales and units of measurement and select the most suitable sampling framework. In this instance, scales and units are relatively straightforward because both sets of measurements (drainage basin and channel) are measurements of area (L^2); in this case km^2 and m^2 for drainage basin and channel areas respectively, are suitable.

Case study: a soil and vegetation survey under coniferous plantations

This study is of a very different nature to the previous study, is much more limited in scope and will be more difficult to use for testing any theories. Can you suggest why this is so? The most obvious difference from the previous case study is that there is no experimental control and therefore no basis for comparison. As a study to demonstrate methods used in soil and vegetation surveys, it is quite adequate but what questions can we answer? Very few, simply because there is no background measurement of other conditions under which soils and vegetation develop. If for example we were comparing the effects of vegetation on soil development, by carefully stratifying the sampling scheme, we could ask many more interesting questions. By choosing our sampling sites in similar positions in the landscape and keeping a constant background geology and climate, it would be possible to compare soil development under conifers and oak for example. This allows us to test ideas as to how important vegetation may be in affecting the development of a soil. If we find no difference in the soil profiles, it may suggest that vegetation has less effect than one of the other variables which we kept constant in our sampling scheme. This then allows us to set up a new series of experiments to test the new idea.

This case study suggests that comparative investigations are more useful for solving problems and that descriptive studies often produce less interesting and unrewarding results.

Case study: meteorological observations

Many of the studies carried out by geographers involve the daily or weekly recording of information. This is the sort of study which is *inductive*, because we start with the data collection before considering how it might be used. As we will see in Chapter 3, if these sorts of data had not been collected by local authority, government or independent organisations in the past it would have severely handicapped what we might have learned about our environment. You may have access to data collected in your school grounds which can form useful background information for understanding weather patterns, especially if national weather forecasts are included with your own data. Once again, however, description alone is not enough when using this type of primary data source. Many interesting studies can be undertaken based around the fixed meteorological station, especially those which supplement the data, perhaps by examining the effects of a local urban area on some of the meteorological variables that you can measure with portable equipment, such as temperature and windspeed.

2.8 CONCLUSIONS

The foregoing sections have established that exercises centred around the use of primary data require careful planning from the earliest stage. Think about the approach you are adopting and the type of analysis you are to perform on the data before you collect it. Remember that your data need to be both accurate and precise. If more than one person is involved in the study, see if you measure things in the same way and get the same results. Above all, avoid the pitfall of suggesting that you have 'proved' so and so. You probably haven't! Finally, remember that the best way to get good data and good results is to be critical of your own work and set up studies which answer questions as well as describe things.

EXERCISE 1:   QUESTIONNAIRE SURVEY

A survey is to be carried out within a village in order to find out whether a recent withdrawal of bus services has caused problems to residents when trying to reach non-local services like doctors, shops and a sports centre. The village consists of 41 households, residing in separate houses numbered 1 to 41. Some characteristics of the resident population are already known, whereas others can only be found out after an interview (Table 2.2).

TABLE 2.2   Household characteristics of village

CHARACTERISTICS KNOWN BEFORE INTERVIEW			CHARACTERISTICS NOT KNOWN BEFORE INTERVIEW				
House No	No of cars in household	Housing tenure OO = owner occupation PS = renting from public sector PR = private renting	Occupation of chief wage earner of household	Size of household	No of children aged <5	Age of other members of household	Cars used by adult members of household to travel to work
1	2	OO	Teacher	4	2	37, 33	1
2	3	OO	Manager with >25 employees	4	—	58, 55, 24, 18	3
3	1	OO	Clerical worker	3	1	32, 31	1
4	1	OO	Chiropodist	4	1	29, 29, 6	1
5	1	OO	Retired	1	—	68	—
6	1	OO	Unemployed	4	—	41, 42, 14, 12	—
7	2	OO	Building soc. manager	4	—	47, 46, 16, 14	1
8	2	OO	Planning engineer	3	1	30, 26	1
9	1	OO	Shoe shop assistant	4	2	31, 30	1
10	—	PS	Farm labourer	6	2	35, 32, 7 6	—
11	—	PS	Farm labourer	5	3	29, 28	—
12	—	PS	Postman	2	—	26, 22	—
13	—	PS	Metal turner	2	—	25, 18	—
14	1	PS	Production line car worker	3	1	26, 24	1
15	2	PS	Unemployed	6	1	40, 38 18, 12, 8	—
16	1	PS	Production line car worker	3	1	28, 24	1
17	1	PS	Welder	4	—	33, 31, 8 6	1
18	1	PS	Production line car worker	4	—	44, 41, 12, 9	1
19	1	PS	Sheet metal worker	4	—	46, 42, 15, 8,	1
20	1	PS	Production line car worker	3	1	27, 24	1
21	2	OO	Bank manager	4	1	40, 36, 7	1
22	2	OO	Dentist	4	1	42, 34, 9	1
23	2	OO	Estate agent	2	—	25, 27	2
24	3	OO	Company accountant	4	—	50, 46 20, 18	4

House No	No of cars in household	Housing tenure OO = owner occupation PS = renting from public sector PR = private renting	Occupation of chief wage earner of household	Size of household	No of children aged <5	Age of other members of household	Cars used by adult members of household to travel to work
25	2	OO	Bookshop proprietor	3	—	58, 57, 29	3
26	2	OO	Teacher	4	2	36, 31	1
27	1	OO	Electrical engineer	6	2	47, 43 13, 11	1
28	1	PS	Schoolteacher	3	2	33	1
29	1	PS	General labourer	2	—	48, 41	1
30	1	PS	Unemployed	2	—	58, 55	—
31	1	PS	Unemployed	3	—	41, 44, 10	—
32	—	PR	Retired	2	—	69, 67	—
33	—	PR	Retired	1	—	82	—
34	—	PR	Retired	1	—	78	—
35	—	PR	Retired	2	—	72, 74	—
36	—	PR	Farm labourer	7	1	44, 42, 15, 11, 9, 7	—
37	2	OO	Solicitor	4	2	38,32	1
38	3	OO	Manager with <20 employees	3	—	47, 41, 9	2
39	2	OO	University lecturer	2	—	58, 51	2
40	2	OO	Clothes shop assistant	4	1	31, 12, 6	1
41	2	OO	Solicitor	3	1	35, 31	2

Q **1** Discuss how you would carry out a random survey of 20 households within the village. Look at the characteristics of the 20 households and consider how representative your sample has been. With reference to the available data what are the strengths and weaknesses of this method of sampling?

2 Devise a 20% stratified, random sample on the basis of the number of cars per household. Why is the number of cars per household likely to be of interest in a survey of this kind? What information not previously known before the interview would cast further light upon mobility problems and why?

3 A systematic survey is to be carried out, interviewing the chief wage-earner of every fifth house (starting with house number 1 and thereafter 5, 10, 15 ... etc). The survey is to be undertaken on Saturday morning. What problems would result if this procedure was carried out?

4 It is thought that different social groups within the village may experience different sorts of problems when trying to reach particular services. A 50% random survey contacted the following houses: 1, 2, 3, 9, 10, 12, 14, 15, 16, 19, 20, 24, 26, 28, 29, 30, 31, 33, 36, 38, 39. Using the available data attempt to sort the population into different social groups according to i) occupation of chief wage earner of household and ii) housing tenure. With reference to the data what other methods of classification may be useful to divide the population into social groups? Justify your choice of categories.

EXERCISE 2:   VEGETATION AND SOIL DEVELOPMENT

This exercise should be attempted in the understanding that the general sampling principles and approaches to environmental problem solving are more important at this stage than either a sound knowledge of statistical methods or of the subjects of biogeography and soils.

i BACKGROUND

The hypothetical region of Britain in Figure 2.6 shows a broad river valley in which is located the small urban centre of Hughtown. This town serves as a route focus for the narrow valley which trends east-west and for the less steep upland region to the south. This centre is also an important market for surrounding farms and forestry plantations. Relief to the north is strong, rising to over 300m at Summit Downs, a windswept and open upland area. Geologically, the central lowland region is covered by alluvium, whilst the middle slopes are directly underlain by shales and slates. The major summits to the north and south of the map are underlain by red sandstone. Large areas of the uplands are used for forestry plantation, whilst in some places are the remnants of a mixed deciduous forest which dominated the area for several hundred years before human occupation. The remaining upland areas are used for grazing on improved pastures whereas the lowland plain is predominantly under cultivation.

Figure 2.6
'Hughtown' and its surroundings

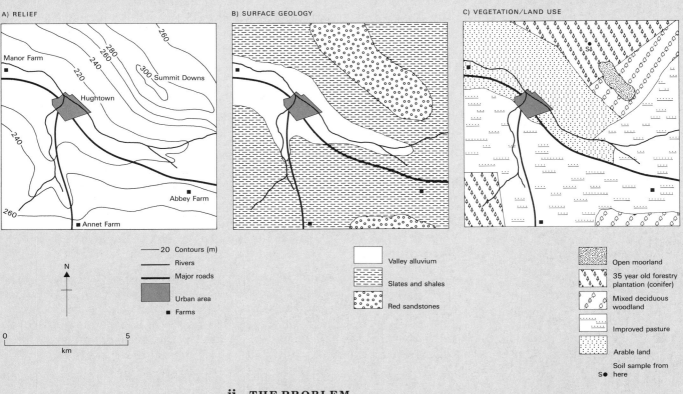

ii THE PROBLEM

TABLE 2.3
Results of soil analysis

The Forestry Commission routinely carries out occasional studies of soil characteristics in its plantations and has been given the results of such a study carried out at the sample point (S) on Figure 2.6. These data are reproduced in Table 2.3.

Horizon	Thickness/Depth	pH
Litter layer	4.0 cm thick	—
Fermentation layer	1.0 cm thick	—
Humus	3.0 cm thick	3.8
A horizon	0–15 cm depth	4.1
B horizon	15–45 cm depth	4.3
B(f) horizon	45–50 cm depth	—
C horizon	50–70 cm depth	4.8

GENERAL INFORMATION.
Soil developed in sandstone on a well-drained site in an area receiving some 965 mm of rainfall per annum. The profile is c 70 cm deep to bedrock and shows evidence of leaching and podsolisation. The pH values are very low, indicating that the soils are acid. These results give some cause for concern, especially where the pH falls below 4 to 4.5.

23

This problem could be caused by a number of factors, but the Forestry Commission have decided that they want to carry out a study to see if the problem is regional or localised and whether it exists only on sandstone or on other rock types. You have been asked to contribute to this study by recommending the sampling strategy that the Commission should adopt in order to see whether:

    i the plantation is changing the acidity of soils in this area;
  ii the local geology is controlling soil acidity.

Your task is to write a report giving answers to the following specific questions provided for your benefit. The questions may be tackled individually or the answers may be given in a more generalised statement of your recommendations. You are NOT expected to carry out the exercise but must give enough information to allow someone else to follow your instructions carefully. You may also wish to request information with regard to the accuracy and precision of the soil analyses already undertaken and should justify any requests.

In order to save money and speed up the study, however, it has been decided that only 100 measurements can be made. Tests will be carried out on samples taken from close to the top of the soil profile (the A horizon). All samples are collected by the same scientist and analysed in the laboratory at the same time.

Q 1 The Commission are interested in more than one factor affecting the acidity of soils and you need to design an experiment to test more than one hypothesis at the same time. Please state the specific hypotheses you wish to test in this study.

2 Having stated the hypotheses to be tested, identify the populations that you need to sample (e.g. how many areas of forest, of what type and of what background geology). Is the whole population available for study or is the target population less than the total population in existence?

3 Explain how you might achieve a degree of experimental control by studying soils in the deciduous woodland as well as in the coniferous plantation.

4 You need to select sampling sites within the area defined on the ground. Explain the relative advantages of using a stratified random sampling scheme in this study (see Chapter 1). Do you wish to sample by area or along transects?

The data you collect are on the ratio scale so that there is little limitation to the type of statistical test you might use to analyse the data. Although important to experimental design, this problem need be of no concern until you have dealt with these topics in Chapter 5. It is only important to remember that the data must not be biased by the methods you use to collect the information.

5 Try to imagine the outcome of the tests you are conducting. Assume that the study shows *no* positive results. Can you formulate other problems which may then have to be investigated to isolate the root cause of the original and still unanswered problem?

Chapter 3 SECONDARY DATA SOURCES

Figure 3.1
Secondary data sources in geography

Secondary geographical data are derived from three major sources as shown in Figure 3.1.

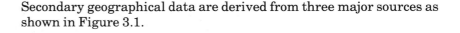

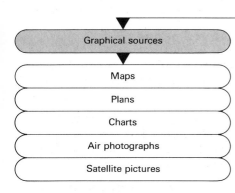

Graphical sources	Written & oral sources	Statistical sources
Maps	Newspapers	Meteorological records
Plans	Geological memoirs	Hydrological records
Charts	Diaries	Land use description
Air photographs	Manuscripts	Census data
Satellite pictures	Radio and television	Industrial statistics
		UN, EEC and UK statistical digests

3.1
MAP SOURCES

Figure 3.2
An Ordnance Survey map on a scale of 1 inch : 1 mile from the OS Old Series (Tavistock) sheet 25

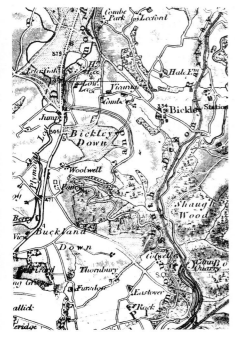

Although this section is later divided into physical and human sources of secondary geographical information, some sources are of value to geographers in general. Of greatest significance are maps. At the present time, maps depict a tremendous range of geographically relevant information at many scales. Most of these are based upon the topographic maps of the Ordnance Survey which, in their own right, form an important source of historical data.

The most familiar contemporary maps published by the Ordnance Survey include the 1:50 000 Topographic, Land Ranger and selected Tourist series. In addition, topographical maps at 1:25 000 and 1:10 560 (now resurveyed at 1:10 000) are available for most areas, and plans at 1:2500 and 1:1250 scale are often to be found for urban areas.

The Ordnance Survey has been publishing topographic maps at the 1:63 360 (1 inch) scale since the early nineteenth century (Figure 3.2), although complete national coverage of the first series was not available until 1873. The second series survey began in some parts of the country in 1840, and the complete survey was published in 1893. Before the introduction of the first series 1:50 000 maps in 1974, seven 1:63 360 series were printed. These various series therefore allow us to make comparative historical studies of settlement, land use, and of coastal and fluvial features at several periods over the last 150 years or so for many parts of the country. Larger scale Ordnance Survey maps, at the 1:25 000 and 1:10 560 date back to around 1840; the first series was completed in 1914 and the second incomplete re-survey was conducted between 1904 and 1923. These larger scale maps are useful for obtaining much more detailed information for some physical features, especially for looking at changes in coastal landforms and river courses through time.

**Figure 3.3
A Tithe Map from the early
nineteenth century**

It is important to remember that the early maps of the Ordnance Survey, and the earlier County Series maps, often used a different projection from the *Transverse Mercator* now in use. Before the First World War a *Cassini* projection was used. Some of these maps are not based on a national meridian (a common origin for starting the map); instead the centre of a county or group of counties often formed the basis for the map projection. The unfortunate consequence of this is that these maps 'distort' towards their margins. As a result we must allow for such distortion on earlier maps in order to compare them directly with modern maps. Since the First World War, however, a Transverse Mercator projection has been used on all Ordnance Survey topographic maps and, at the same time, the familiar grid of 1 km^2 was superimposed over the UK. This system allows us to define very precise locations on maps and often forms a useful basis for the objective sampling of land use or other data. (A detailed discussion of the early history of the Ordnance Survey and of the types of map and projection used is to be found in J B Harley (1975) Ordnance Survey Maps, published by the Southampton Ordnance Survey).

Many other historical map sources are available to the geographer. One of the more useful is the Tithe Survey. This was an exacting survey, mainly covering rural areas, and initiated by an Act in 1836. In addition to the existence of large scale maps, the *Apportionment Books* associated with this survey give land use and ownership on a field-by-field basis. The survey covered some 75 per cent of the UK and continued until 1840 (Figure 3.3).

Most major county and city reference libraries will contain a small selection of maps of various scales and types, but for your own local area the most important source of information is the County Records Office (CRO) which exists for all counties and principal cities. These offices are run by qualified archivists with good knowledge of the local area. Many of these offices contain catalogues and indexes of available material and, if historical maps are required, the CRO should be your first point of enquiry. At a national scale, catalogues, indexes and some of the original field surveys of the Ordnance Survey are housed in the British Library and the British Museum in London and in the Bodleian Library in Oxford.

A wide range of maps depicting various aspects of the physical and human landscape have been published, and in the post-Second World War period the extent of specialist maps has expanded considerably. The application of these maps is discussed in the appropriate sections but an important specialist source of use to both human and physical geographers is the Land Utilisation Map. The First Land Utilisation Survey of Great Britain, directed by Sir Dudley Stamp, began in 1930, published its first 1:63 360 Land Use Map in 1933, and was completed by 1946. These maps, along with a series of handbooks, form an important record of land use before the changes which took place during the Second World War. The Second Land Utilisation Survey, directed by Alice Coleman from 1960, is more detailed and is published at the 1:25 000 scale.

3.2
INTERACTIVE VIDEO

The most exciting secondary geographical data source of recent times is the BBC's *Domesday Project*. This attempts to present in several million pages of information a detailed portrait of the UK, its environment, people, society and organisations in the late twentieth century. Apart from the depth and range of data on offer, what further distinguishes the project is its use of a microcomputer to retrieve data stored on an interactive videodisc system. If printed, the results of the Domesday Project would fill over 300 volumes. Instead, one disc will store over 400 megabytes of digital data (that is, equivalent to 400 million alphabetic characters). The discs are 'read' by laser and can be quickly and simply accessed by using a keyboard or tracker ball. Through video technology the user has immediate access to maps, photographs, aerial views, descriptions, statistics, charts and moving video sequences.

The initial Domesday package consists of two discs, termed local and national. The *local or community disc* consists of some 150 000 screen

26

pages of text and 80 000 pictures of places and landscapes. Data for this disc were collected by over 15 500 schools and community groups, each responsible for a 4 × 3 kilometre block. The organisation of the disc is based on some 24 000 Ordnance Survey maps arranged on six levels:

Level 0 – The United Kingdom: satellite photograph, text.
Level 1 – Countries and Island Groups: satellite photograph, text.
Level 2 – 40 × 30 km regions: satellite and aerial photographs, maps and text.
Level 3 – 4 × 3 km local blocks: community photographs, maps and text.
Level 4 – Street maps, special feature photographs, text.
Level 5 – Floor plans of special sites, special feature photographs, text.

This combination of OS maps, photographs and locality-specific text is nothing less than a complete data base for regional geography.

The data contained on the local disc can be displayed only for local areas, mostly by means of maps. By contrast the *national disc* is a very wide data base. It represents vast quantities of information drawn from official and public sources, articles and reports. The information is organised into four main groups:

Culture including arts, beliefs, language, leisure, religion, sport, customs, fashion, media, crafts, festivals.
Economy including personal finance, national economy, public sector finances, labour relations, prices, consumptions.
Society including education, health, housing, defence, welfare, people, events, transport, communications, law and order.
Environment including conservation, climate, agriculture, ecology, water resources, urban environments, wildlife, landscape, oceanography, energy.

In all there are some 9000 sets of data and 40 000 photographs on the national disc, arranged for rapid access through a simple system of keywords and to a more limited extent, through maps. The data are available by whatever geographical areas were used to report them: these areas include grid squares, parliamentary constituencies, districts, counties, functional regions, Local Education Authority areas, Regional Health Authority areas, television regions and Standard Regions. Postcode sectors and wards may also be available. Wherever possible the same data will be stored for all areal units for which they can be assembled.

In comprehensiveness and ease of use the Domesday Project provides a unique data source. As a basis for general information about the United Kingdom in the 1980s and as a linkage between maps and pictures it has much to commend it. Many projects can be designed around the information stored on these discs. In time it is hoped that many schools will have direct access to the Domesday package, although it is likely that County Records Offices, libraries, universities, polytechnics and colleges will be among the first to acquire the system.

Ecodisc is an interactive video system which brings together a vast collection of data in the form of photographs, graphic displays, film sequences, and printed information about a real nature reserve, Slapton Ley in South Devon. The purpose of the disc is to allow you to experience the reserve: to see what it looks like, to find out what is there and to consider some of the problems associated with managing an ecologically complex area. In your role as Nature Reserve Manager you can view, walk around, sample and watch the reserve and when ready you can formulate a plan concerned with the reserve's future. The videodisc will present your ideas to a set of interested parties who will respond to your suggestions. It is up to you either to change your plans or to carry them out. Ecodisc is available from BBC Enterprises. The *Countryside* disc is a similar and recent addition to the list.

Secondary data sources in human geography

Much data in human geography are derived from official statistics. A number of general guides provide information on where certain forms of data may be obtained. For international sources, useful digests include the *Directory of International Statistics* (United Nations, 1975) and Wasserman's (1977) *Statistical Sources*. Details of the publications of the Statistical Office of the European Community are found in Ramsay's (1986) *Guide to Official Statistics*. Maunder (1974–) has supervised a series of *Reviews of United Kingdom Statistical Sources*, with volumes covering personal social services, health, housing, tourism, transport, land use and energy, among others. Current developments in British official statistics are recorded in the quarterly journal *Statistical News* and in the Central Statistical Office's *Guide to Official Statistics*.

Some national newspapers provide important sources of general information relevant to geographers. *The Guardian, Daily Telegraph, Independent, Financial Times* and *The Times* all contain occasional reports on policies, planning proposals and issues relevant to an understanding of contemporary society.

3.3 SECONDARY DATA SOURCES FOR POPULATION GEOGRAPHY

Historical sources: changes in time

Studies in historical geography are often concerned with either reproducing population patterns in the past or observing changing population distributions over time.

A starting point for studies of both kinds are *Parish Registers*. These are documents which record christenings, weddings and burials in 11 000 English parishes between 1538 and 1837, when civil registration began. Most County Record Offices have large collections of parish registers and these are listed in *Parish Registers in Records Offices and Libraries* (1974–).

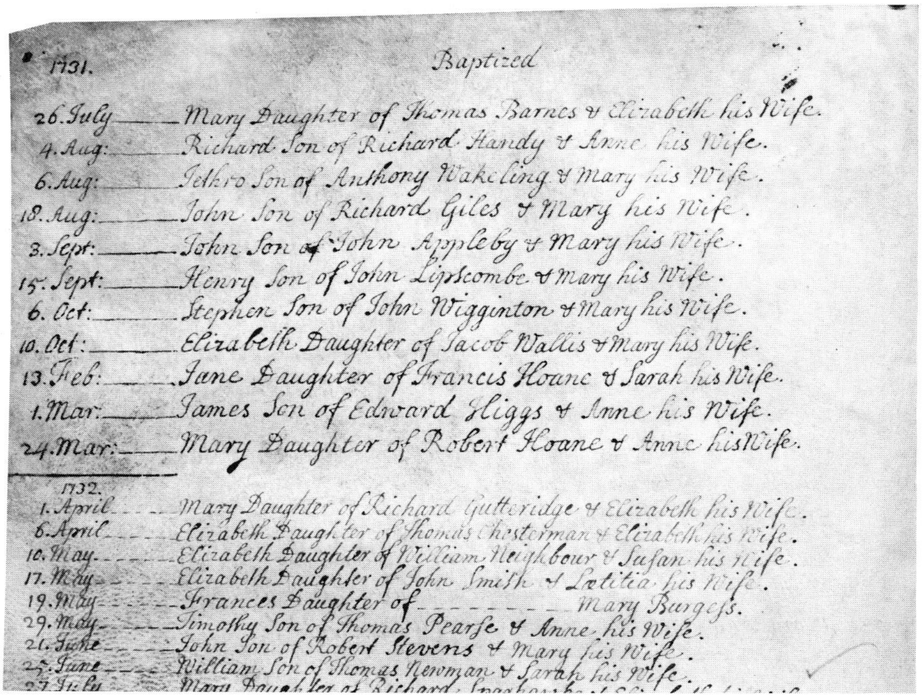

1731. Baptized
26. July — Mary Daughter of Thomas Barnes & Elizabeth his Wife.
4. Aug: — Richard Son of Richard Handy & Anne his Wife.
6. Aug: — Jethro Son of Anthony Wakeling & Mary his Wife.
18. Aug: — John Son of Richard Giles & Mary his Wife.
3. Sept: — John Son of John Appleby & Mary his Wife.
15. Sept: — Henry Son of John Lipscombe & Mary his Wife.
6. Oct: — Stephen Son of John Wigginton & Mary his Wife.
10. Oct: — Elizabeth Daughter of Jacob Wallis & Mary his Wife.
13. Feb: — Jane Daughter of Francis Hoane & Sarah his Wife.
1. Mar: — James Son of Edward Higgs & Anne his Wife.
24. Mar: — Mary Daughter of Robert Hoane & Anne his Wife.
1732.
1. April — Mary Daughter of Richard Gutteridge & Elizabeth his Wife.
6. April — Elizabeth Daughter of Thomas Chesterman & Elizabeth his Wife.
10. May — Elizabeth Daughter of William Neighbour & Susan his Wife.
17. May — Elizabeth Daughter of John Smith & Letitia his Wife.
19. May — Frances Daughter of ------ Mary Burgess.
29. May — Timothy Son of Thomas Pearse & Anne his Wife.
21. June — John Son of Robert Stevens & Mary his Wife.
25. June — William Son of Thomas Newman & Sarah his Wife.

Figure 3.4
An extract from Parish Register

Because of their wide coverage, parish registers provide the main source material from which the historical geography of the population of England may be traced. It is a relatively straightforward task to compare population characteristics of different parishes and regions, or to study a parish over time. Many early modern towns contained a large number of small parishes which enable contrasts to be drawn between different areas in cities.

By carefully sifting through parish registers a range of population characteristics can be noted:

i *Population trends:* The simplest form of analysis is to make annual or monthly tabulations of the frequency of baptism, marriage and burial entries.

ii *Demographic rates:* Infant mortality rates are calculated by noting the number of deaths occurring within a year from birth per thousand live births. As ages at death are sometimes given these can be used to work out average life expectancy. Other rates include estimates of fertility, mortality and illegitimacy.

iii *Migration:* After 1753 place of residence of marriage partners is recorded. By looking at marriage registers assessments of migration and marriage distances are made.

iv *Family reconstruction:* Detailed inspection of parish registers through time provides some idea of local family histories.

TABLE 3.1
Historical sources of population data

Problems with parish registers include their survival and accuracy. A conventional viewpoint is that parish registration was relatively accurate in the early eighteenth century, became somewhat less so in the 1780s, virtually collapsed between 1795 and 1820, and then improved between 1821 and 1837.

Before the period when Census data provide a reliable source, a wide range of other evidence has been used to study population. Hollingsworth (1976) lists eighteen, ranging from vital registration records to tombstone inscriptions in cemeteries. Often these sources provide additional information on social and economic conditions. Table 3.1 provides a summary of some of those most useful for project work held by County Record Offices and the Public Records Office.

Topic	Source	Period	Data
Population	Poll taxes (e.g. Subsidy Roll of Edward III)	1377	Population of England, except Cheshire and Durham and children under 14
	Parish registers	1538–1837	Christenings, marriages, burials
	Ecclesiastical censuses (e.g. Compton Census)	1676	A count of communicants, Catholic & Non-Conformists in parishes, aged over 16
	Bills of mortality	Various	Deaths
	Civil Registration records	1837–	Births, marriages, deaths
Wealth	Lay Subsidy surveys	1524–5 1543–5	Tax on personal property, landed income & wages. Includes all those with income greater than £1/annum except those in Holy Orders and children under 16
	Hearth Tax	1662–89	Tax on hearths. Informs on households; poor were exempt
Poverty	Poor Law returns: Old Poor Law New Poor Law	1601–1834 1834–1890*	Indoor and outdoor relief given to the poor
Literacy	Protestation Oath Test Oath Wills Parish registers	1642 1723 various 1538–1837	Those unable to sign other than by a mark
Occupations	Enrolment into apprenticeships	Various	Trades
	Records of admission to town freedom	Various	Trades and manufacturers
	Trade directories	Various	Trades and manufacturers

* After 1890 less useful for highlighting patterns of poverty.

From Table 3.1 two sources are of especial interest. The *Hearth Tax* of 1662 established a level of two shillings for every hearth, unless the occupant was exempt on the grounds of poverty. Returns have survived for many areas. These are widely used as they inform not only about the distribution of wealthy households but also, indirectly, about populations. In order to estimate population, the number of households (that is, number of hearths) is usually multiplied by 4·5, which is a value based on an estimate of average family size.

The *Annual Returns of the Poor Law Guardians* provide a rich source for geographical studies of poverty. From 1601 the payment of *outdoor relief* to those in need was administered by Boards of Guardians within parishes. The system continued until 1834 when an Amendment Act tightened up Poor Law administration. Records of payment have survived for many parts of England and Wales. Their attractiveness is the range of information they contain (Figure 3.5).

Figure 3.5
An annual return of the Poor Law Guardians

Using the Poor Law returns and in conjunction with the Post Office Directory for 1845 it was possible to reconstruct the pattern of street numbering and thus, to establish where those on the 1847 list lived (Figure 3.6).

Figure 3.6
Out-relief recipients in central Birmingham 1847

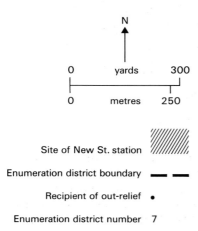

Site of New St. station

Enumeration district boundary

Recipient of out-relief •

Enumeration district number 7

The main drawback with these returns is that not all those who were poor received out-relief. By the 1890s the discrepancy between recipients and the total number of poor was so great that the records are of little significance in highlighting patterns of poverty.

A watershed in the collection of population statistics came with the first national *Census of Population* in 1801. Since then Censuses have been carried out in the first year of every decade, the only break coming in 1941. The Censuses of 1801 and 1831 suffer from large errors, since the data collection was spread over a number of days which increased the possibility of double counting. From 1841 onwards, Census counts were completed in one day. The Censuses of 1841 to 1881 provide especially valuable information. Although increasing numbers of house-by-house records associated with the Censuses of the early nineteenth century have been discovered in recent years, 1841 was the first year in which Census schedules were completed for every household and later copied into enumeration books. Under a hundred-year confidentiality rule, 1881 is the most recent Census for which such detailed data are currently available for England and Wales. Census enumerators were persons responsible for collecting information on defined areas. Large urban centres were divided into many enumeration districts (Figure 3.7)

Figure 3.7
Parish of Birmingham in 1851
Census Enumerators' districts

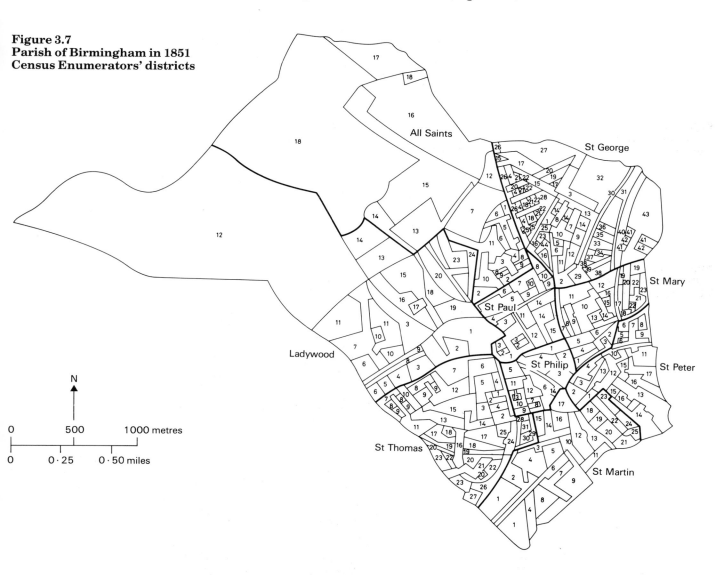

Census enumerators' books can be inspected in the Public Records Office, although some of the larger regional reference libraries store this information usually on microfilm. The returns provide details on individuals, their households and the areas in which they lived. Many projects can be designed around the Census. Of particular interest are studies of the population geography of certain areas. By plotting the distribution of selected characteristics patterns can be described and comparisons made with various urban models.

Contemporary sources: changes in space

There are many contemporary sources which inform about aspects of population geography. Of these the *Returns of the Registrar General* provide the most detailed inventories of the demography of England and Wales and the UK. The information is collated by the OPCS and published in a variety of forms. These include the *Registrar General's Annual Estimate of Populations in England and Wales* and the *Registrar General's Quarterly Returns for England and Wales* (1966–73) now called *Population Trends*. These sources are especially valuable for geographical study as data are presented by country, Standard Region, and Health Region, and by age and sex. International migration records distinguish country of origin. Tables cover population, population change, vital statistics, births, marriages, divorces, migration, deaths and abortions.

These returns are also useful for studies of medical geography. For

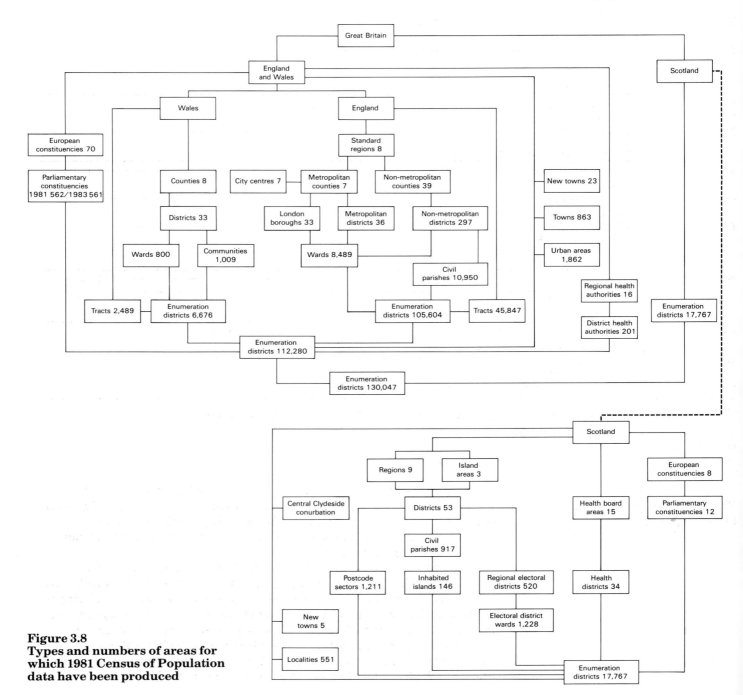

Figure 3.8
Types and numbers of areas for which 1981 Census of Population data have been produced

32

Country	Topic	AGGREGATION LEVEL			
		Country level	Standard region	Counties	Local Authority Areas (Districts)
	Birth place & usual residence				
GB	Country of birth	X	X		
SC	Country of birth	X			
GB	Usual residence				X
	Migration				
GB	National migration	X	X	X	
SC	Migration report			X	X
NI	Migration report	X			
EW	Regional migration		X	X	X
	Housing, households and families				
EW	Housing & households	X	X		
SC	Housing & houshold reports	X			X
EW	Household & family composition	X	X		
SC	Household & family composition	X			
GB	Communal establishments	X	X		
	Economic activity				
GB	Economic activity	X	X		
SC	Economic activity	X			
NI	Economic activity	X			X
	Workplace and transport to work				
EW	Workplace and transport to work				X
SC	Workplace and transport to work				X
NI	Workplace and transport to work				X
	Language				
SC	Gaelic report				X (1)
WA	Welsh language				X (2)
	Others				
GB	Sex, age and marital status	X		X	
GB	Persons of pensionable age	X	X	X	
GB	Qualified manpower	X	X		

GB Great Britain   SC Scotland   EW England and Wales   WA Wales   NI Northern Ireland

(1) also for wards/parishes, postcodes, Special Areas.

(2) wards/parishes.

TABLE 3.2
Special topic volumes of 1981 Census of Population

example, the section on deaths reports the cause of death tabulated by Health Region. Further regional health data are available in the *Registrar General's Weekly Return for England and Wales,* which reports on notifications of infectious diseases in Standard Regions, administration areas and Port Health Authorities.

Another principal source is the national *Census of Population*. The Censuses of 1961, 1971 and 1981 offer the most comprehensive insight into contemporary society. These Censuses cover the whole population living in the UK on Census night. Even though the Census collects a wide range of information about individual people, it is not concerned with people as individuals. The Census is taken solely to compile statistics about groups, categories and in particular, geographical areas. The great variety of areal units for which Census data are available is shown in Figure 3.8.

Information is also available for National Grid squares. These have the advantage of being regular and permanent. Their disadvantage is that they do not correspond with any physical feature or administrative area.

In 1981 data were categorised into 39 tables, some of which were based on a 10% sample of returns in order to reduce costs. The main topics include:

Population numbers	Occupation
Sex composition	Employment status
Age structure	Educational qualifications
Marital status	Social class
Fertility	Method of travelling to work
Population movement	Ownership of motor vehicle
Country of birth	Housing tenure
Household structure	Housing amenities and space

Special topic volumes have also been published (Table 3.2).

Censuses have varied in their presentations and data collection. In consequence, comparison of areas over time is difficult. This can be illustrated by looking briefly at some differences between the 1971 and 1981 Censuses:

i *Boundary changes.* In 1974 Local Government reorganisation led to many boundary changes and these were recognised by the Census;

ii *Range of questions.* In 1971 the number of questions asked was 29 (5 household and 24 population) in 1981 this was reduced to 21 (5 household and 16 population);

iii *Types of question.* In 1981 some questions on household amenity were dropped and less detail was gathered on ethnicity (no questions of date of entry into UK, parents' birthplaces, language, religion or other 'cultural' characteristics).

For England and Wales the Census is available from Census Customer Services, OPCS, Titchfield, Fareham, Hants PO15 5RR, and for Scotland from Census Customer Services, GRO (Scotland), Ladywell House, Ladywell Road, Edinburgh EH12 7TF.

Three official surveys help fill in some of the information gaps about social changes between Censuses.

The *General Household Survey* (GHS) is a continuous survey based on a sample of the general population resident in private (non-institutional) households in Britain and has been running since 1971. In recent years about 12 000 households participated in the survey. It covers aspects of population and fertility, housing, employment, education and health. Both annual statistics and general trends are shown. Certain subjects are covered periodically; for example, the 1983 GHS included a section on leisure activities, a topic previously presented in 1973, 1977 and 1980. Topics are introduced when of public concern; recent additions include contraception and ethnicity. A major limitation for geographical study is its lack of regional statistics. However, the Social Surveys Division of the OPCs frequently produce supplementary reports and monitors which highlight the regional dimensions of its data base.

The *Labour Force Survey* (LFS) is a large-scale household survey (57 000 households in 1984) carried out every two years until 1983 and from 1984 carried out annually. The report provides information on economic activity, country of birth, ethnic origin and nationality, labour force characteristics and housing circumstances. Some of the data are provided by Standard Region.

A further official source is *The Family Expenditure Survey*. This is an annual survey conducted by the Department of the Environment. It is based on a sample of 11 000 private households in the UK and has been continuous since 1957. It represents a reliable source of household data on expenditure, income and other aspects of household finances. The survey provides a perspective of national changes in such matters as spending on items as diverse as food, clothes, fuel and alcohol over more than a quarter of a century. Much of the data are tabulated by Standard Region, allowing, for example, comparisons between income in the South East and Northern Ireland or illustrating differences in diet between Wales and Scotland.

Sources for the study of resources and levels of development

Table 3.3 illustrates selected sources available at an international level. In most cases these are summary statistics compiled by individual countries. The United Nations (UN), the Organisation for Economic Co-operation and Development (OECD) and Eurostat (European Community) publish many other specialised accounts. In addition, other organisations of countries supply their own particular statistics, such as EFTA, COMECON and OPEC. Since 1979 the World Bank has published a useful statistical review in its World Development Report. These sources are useful for comparative studies of national or regional characteristics, although some care needs to be taken when interpreting the results. Different countries may well have used different criteria for classification. For example, there is no universally agreed way of defining 'urban' and not every country uses the Standard Industrial Classification employed by the UK and USA.

Member countries of OECD: Austria, Belgium, Canada, Denmark, France, Federal Republic of Germany, Greece, Iceland, Ireland, Italy, Luxembourg, Netherlands, Norway, Portugal, Spain, Sweden, Switzerland, Turkey, United Kingdom, USA (all 1961), Japan (1964), Finland (1969), Australia (1971), New Zealand (1973).

Member countries of EC: Belgium, France, Italy, Luxembourg, Netherlands, Federal Republic of Germany (all 1957), Denmark, Ireland, United Kingdom (all 1973), Greece (1981), Portugal and Spain (1986).

TABLE 3.3   Selected international data sources

Topics	Worldwide	Publication	European Comm. (Eurostat)	Publication
General	UN Statistical Yearbook	Annually 1949–	Basic Statistics of the Community	Monthly 1958–
	UN Bulletin of Statistics	Monthly 1947–	Eurostat Review (10 year development review)	Annually 1981–
	Statesman's Yearbook	Annually 1964–		
	Europa Yearbook	Annually 1926–	Yearbook of Regional Statistics	Annually 1981–
Population & social conditions	UN Demographic Yearbook	Annually 1948-	Demographic Statistics	Annually 1977–
	WHO World Health Statistics	Annually 1951–		
	UNESCO Statistical Yearbook	Annually 1963–		
Agriculture	UN Yearbook of Agricultural Statistics	Annually 1947–	Yearbook of Agricultural Statistics	Annually 1970–
Trade, industry & economic affairs	UN Yearbook of National Account Statistics	Annually 1958–	Eurostatistics: data for short-term economic analysis	Monthly 1979–
	UN Industrial Yearbook of Statistics	Annually 1974–	Yearbook of Industrial Statistics	Annually 1984–
	UN Yearbook of International Statistics	Annually 1951–	External Trade Statistics	Annually 1976–
	OECD Main Economic Indicators	Monthly 1965–		
	OECD National Accounts	Annually 1967–		
	OECD Labour Force Statistics	Annually 1961–		
	OECD Statistics of Foreign Trade	Annually 1974–		

3.4
SECONDARY DATA
SOURCES FOR ECONOMIC
GEOGRAPHY

Sources for agricultural geography

Studies of agriculture and land use offer rewarding topics for project work. Two principal sources of secondary data are available.

Land use surveys and classifications: The value of these surveys is that they provide a broad description of farming patterns and landscapes and, when compared, provide an assessment of agricultural change.

In Britain, two national Land Utilisation Surveys have been undertaken. The *First Land Utilisation Survey* began in 1930 and was based on a simple sixfold division of land use. It was published as a series of one inch maps by the Ordnance Survey.

The *Second Land Utilisation Survey* was started in 1960 and divided land into twelve groups. Maps were published at 1:25 000 scale. In addition a second set grouped land into five categories: wildscape, farmscape, townscape, marginal fringe and urban fringe. These maps were printed at 1:400 000 scale.

Type of farming classifications, which consider livestock and crop characteristics, are also available. In 1967 the Ministry of Agriculture, Fisheries and Food (MAFF) classified England and Wales into six farming types. This was based on a one-sixth sample of individual holdings and the results were mapped using $10km^2$ grids. The following year MAFF produced separate maps for each farm type. In the early 1970s two other forms of map were introduced at a scale of 1:250 000 and produced for each of their eight Ministry regions: type of farm by size of farm and type of farm by size of farm business. Data are shown for individual farms in the form of colour coded proportional circles.

The problems of using classifications include the following:

i comparison is often difficult as land use categories are rarely the same in different surveys;

ii classifications are man-made and detail will vary according to such things as scale and cut-off points between classes.

Agricultural census: The agricultural census provides information on crop acreages, livestock numbers, farm types and sizes, the number, sex and type of agricultural workers and tenure. The census has five main uses for project work:

i to show spatial variation in farming over wide areas;

ii to provide the basis for agricultural classifications;

iii to consider the evolution of farming patterns;

iv to study the spread and diffusion of new developments, such as oil-seed rape, over time and space;

v to act as a sampling framework for farm-based questionnaires.

Data have been collected annually on most agricultural holdings throughout the United Kingdom since 1866. On 4th June each year farmers are obliged to complete a return on their farm, providing it is larger than 4 ha (10 acres) in size. For confidential reasons information is grouped into larger administrative units when published.

At the *regional* and *county* level statistics are presented in the annual publications of the MAFF (England and Wales), the Department of Agriculture and Fisheries for Scotland and the Department of Agriculture for Northern Ireland. The results are also produced at a *parish* level, which is the smallest agricultural administrative unit (Figure 3.9). Parish summaries for England and Wales are held at the Public Records Office and regionally at the local offices of the Agricultural Development Advisory Service (ADAS). In Scotland parish data are found at the Scottish Record Office, Edinburgh.

The main problems of using the agricultural census are:

i parishes have declined in number over time, making comparison difficult;

Item No.	CROPS, GRASS AND LABOUR	Item No.	FARMERS AND WORKERS	Item No.	CATTLE	Item No.	PIGS	Item No.	VEGETABLES (grown in the open)
1	TOTAL AREA			70	Cows and heifers in milk — mainly for the dairy herd	100	Sows in pig	170	Brussels sprouts — f m.
2	LAND RENTED	50	Farmers or partners — W t — Working on the holding	71	mainly for the beef herd	101	Gilts in pig	171	f p.
3	LAND OWNED	51	P t	72	Cows in calf but not milk — for the dairy herd	102	Breeding pigs — Other sows	172	Cabbage, summer and autumn
4	TOTAL CROPS AND FALLOW	52	Wives or husbands of principal farmers or partners (working on the holding)	73	for beef herd	103	Boars for service	173	All other Cabbage
5	Grassland (1980 or later)	53	Other partners and directors — W t — Working on the holding	74	Heifers in calf (first calf) — mainly for the dairy herd	104	Gilts 50 kg (110 lb) and over	174	Cauliflower summer and autumn
6	Other Grassland (excl. Rough Grazing)	54	P t	76	for calves in the beef herd	105	Barren sows for fattening	176	Calabrese (sprouting broccoli)
7	Rough Grazing	55	Wives or husbands of other partners and directors (working on the holding)	78	Bulls for service — 2 years old & over		All other Pigs:	178	Carrots
8	Woodland	56	Salaried managers	79	1 year old & under 2	106	110 kg and over (240 lb and over)	181	Parsnips
9	All other land	57	Other Family Workers — Regular W t — Male		All other Cattle and Calves:	107	80 kg - 110 kg (175 lb - 240 lb)	183	Beetroot — f m
		58	Female	80	2 years old and over — Male (excluding bulls for service)	108	Pigs weighing (live weight) — 50 kg 80 kg (110 lb - 175 lb)	184	f p
	CROPS AND FALLOW	59	Regular P t — Male	81	Female — for slaughter	109	20 kg 50 kg (45 lb 110 lb)	185	Onions — for salad
11	Wheat f.t	60	Female	82	others	110	Under 20 kg (45 lb)	186	f h d
33	Durum Wheat	61	Hired Workers — Regular W t — Male	83	Male (excluding bulls for service)	111	TOTAL PIGS	187	Broad Beans
12	Barley – Winter f.t.	62	Female	84	1 year old and under 2 — Female — for slaughter		SHEEP	189	Runner Beans (pinched)
13	Barley – Spring f.t.	63	Regular P t — Male	85	for the dairy herd	113	Ewes for breeding	190	Runner Beans (climbing)
14	Oats f.t.	64	Female	86	for beef herd	114	Two-tooth ewes	192	French Beans
15	Mixed Corn f.t.	65	Seasonal or Casual workers — Male	87	6 months old and under 1 year — Male (including bull calves for service)	115	Rams for service	194	Peas f h d
16	Rye f.t.	66	Female	88	Female	116	Draft and cast ewes	196	Green peas — f.m.
17	Maize f.t. or f.s.f.	69	TOTAL FARMERS AND WORKERS			117	Wethers and other sheep	196	f.p.
19	Potatoes (early and maincrop)	48	Youth Training Scheme Trainees			118	Lambs under 1 year	197	Celery (not under glass)
20	Sugar Beet not f s f					119	TOTAL SHEEP AND LAMBS	198	Lettuce (not under glass)
21	Hops						POULTRY	199	Sweet Corn
22	Horticultural Crops (excluding mushrooms)							200	Other vegetables and mixed areas (incl. sweetcorn and watercress)
23	Beans							201	TOTAL VEGETABLES GROWN IN THE OPEN
24	Turnips and Swedes								
25	Crops for stock — Fodder Beet and Mangolds								

Figure 3.9
An extract from the Guide Key to parish summaries

ii by holding the census on 4th June certain crops may go unreported. For example, some crops may have already been harvested and others not yet planted;

iii definitions of what constitutes a farm have varied. Between 1892 and 1968 returns were compulsory for all farms of more than 1 acre in size. The latest definition (>4ha) led to the exclusion of some 47 000 holdings;

iv there is some evidence to suggest that entries by some farmers are inaccurate, whether through ignorance, uncooperation or bias, although such errors are small in relation to the total.

Sources for industrial geography
Employment and production: The Censuses of Population, Employment and Production provide the most important official aggregate statistics for industrial geography. These sources are useful for a number of reasons:

i ready availability;

ii most series cover the whole population or, at least, a very large sample of individuals or firms;

iii national coverage, enabling regional comparisons;

iv data are collected and published at regular intervals, permitting change over time to be assessed.

The *Census of Population* is the main source of information on the occupational structure of the workforce, the number of self-employed and details of journey to work. Information is available on those working full-time, part-time, seeking work and retired. The Census distinguishes between the type of job done and the type of establishment in which the work is done (activity headings of the SIC). The results are published in topic volumes.

TABLE 3.4
Extracts from the standard tables
of the County Reports

The County Reports of England and Wales provide additional tabulations, some of which are outlined in the table below.

Table No.	Base Population	Characteristics & Cross Tabulations
	Economic characteristics	
12	Pop. usually resident aged 16+	Economic position × age × marital status × sex
13	Pop. usually resident aged 16+ in employment	Employment status × sex × marital status (women)
14	Pop. usually resident aged 16+ in private households	Economic position × marital status × sex
15	Pop. usually resident aged 16–24 in private households	Economic position × marital status × age (single years) × sex
16	Married females usually resident in private households	Economic position × age
	10 per cent tables	
43	Pop. usually resident 16+ in employment	Industry × age × working outside district of residence × sex
44	Pop. usually resident 16+ in employment	Socio-economic group × working outside district of residence × working full-time, part-time
45	Persons aged 16+ usually resident in private households	Cars per household × means of travel to work × working full-time, part-time × outside district of residence
46	Pop. usually resident economically active or retired	Socio-economic group × economic position × sex × marital status (women)
47	Private households with usual residents	Socio-economic group of active and retired head, never active head × tenure of households in permanent buildings, with no car; retired heads × socio-economic group

Some of the main problems of using the Census as a source of economic data include the following:

i the data available for Northern Ireland and Scotland often differs from that of England and Wales;

ii the data are presented in different forms in different Censuses. In 1971 the economic activity tables gave employment by workplace data only for large local authorities, details for small urban administrations were not provided. In 1981 the Census gave these data for all local authority districts. Also, a different industrial classification was used in 1971 than that in 1981;

iii slow publication of data. Economic activity tables were not published until 1983.

A more frequently used source for employment data is the *Census of Employment (CE)*. This collects information about 'employees in employment' during the week of the Census. This was held annually in June from 1971 to 1978, since then it has been taken every three years, the latest in 1984. Returns are completed by the employer. The results are published in the *Employment Gazette*. Tables show employment levels in each Standard Region of Great Britain by activity headings of the Standard Industrial Classification.

The value of the CE is that it provides a good guide to regional trends in employment. Its weakness are:

i it says nothing about which firms are going in and out of business;

ii definitions of employer have varied. In 1984 the survey covered all employers with 10 or more employees, only a 10% sample was taken of the rest;

iii two job workers are counted twice, self-employed are missed and so too are part-time workers not working on the Census week.

The Employment Gazette also publishes monthly national and quarterly regional estimates of employment, as well as a monthly up-date of those registered as unemployed at job centres. Unemployment statistics are presented for Assisted Areas, Employment Office Areas, travel-to-work areas and counties.

An additional source of employment data is the *Labour Force Survey*. This includes all persons who did some paid work in the reference week and so covers the self-employed, omitted by the CE.

The *Census of Production* provides a record of national and regional trends in industrial development. Information includes output, expenditure, stocks, sales, employment, wages and work in progress. Up until 1968 the Census was held every five years; since 1970 it has become an annual Census. The results are reported in issues of the *Business Monitor*. A separate report is published on each industry. Until 1979 these statistics were available for regions, but since 1980 regional results only show broad groupings of industries. Some of the most useful tables are the numbers of manufacturing units in various size groups, by industrial classification and by area, together with the total number of persons employed in each category.

There are some general problems which relate to all three Censuses:

  i the data are collected to serve the needs of government and so may not be in the form most suited to project work;
 ii the timings of the different surveys often do not coincide, making comparison difficult;
iii different counting and classification methods are used by each Census;
 iv there is a general lack of small area data. With the exception of the Census of Population, most data are only broken down to Standard Regions.

Firm and establishment data: Information on individual enterprises are valuable for a number of reasons:

  i industrial patterns within small areas can be mapped, such as intra-urban locations;
 ii *components of change* can be identified, that is, the number of industrial births, deaths, movers and stayers within a local area;
iii data can be combined into categories suitable for project work.

The are a number of more accessible data sources. Trade directories are useful as historical sources, some dating back to the late seventeenth and early eighteenth century, and for up-to-date inventories. Many different types exist, but most can be grouped as general directories of particular areas or specialist directories of specific trades.

Kelly's Directories are some of the most useful general street directories. Information is provided on the name, address and nature of the manufacturing or retail function of an establishment. From 1845 to 1946 Kelly's Directories were published for most counties of the United Kingdom and for most of the largest towns and cities. From 1946 to 1976 Kelly's published annual directories for large towns and cities only. Since 1976 Kelly's has ceased publication except for their *Post Office London Directory*. Most reference libraries keep copies of these directories, although the earliest publications are most likely to be stored in the local archives office.

Telephone directories, both yellow and white page Post Office directories, are especially useful for checking dates of openings and closures of enterprises and for identifying local movement.

Additional information on ownership and status can be added by reference to *Who Owns Whom*, published annually since 1958, and *Kompass UK*, also published annually.

Supplementary sources which may be available in local commercial libraries or larger reference libraries are the *Industrial Market Location Directory* and Dun and Bradstreet's *Key British Enterprises*. The former consists of two volumes for each county. Map pages identify the exact location of manufacturing plants, wholesalers and vacant premises in the main industrial areas. Further tabulations include company name, address, male and female employment, status of plant ownership and industrial type. The latter provides a two-volume listing of companies, with a brief description including employment, location and activities.

Common problems with firm and establishment data include coverage, accuracy and comparability:

i not every enterprise in an area will be included, inevitably very small plants will often be missed because they fail to register or respond to surveys;

ii some enterprises will have been misclassified because of poor information supplied by owners;

iii different directories categorise plants in different ways.

Sources for the geography of transport and trade

Projects in transport geography include: i) assessments of inter-regional linkages; ii) comparisons of regional levels of accessibility; iii) studies of the characteristics of transport networks; iv) evaluations of the comparative importance of modes of transport. A number of data sources are useful in these respects. The most important and extensive record of transport information is provided in *Transport Statistics of Great Britain* (Department of Transport, Welsh Office, Scottish Development Department). An annual publication, extending back for some twenty years, it presents a comprehensive account of air, sea, inland waterway, rail, pipeline and public and private road transport. Tables relate to goods, passengers and routes, with additional regional coverage.

Each mode of transport can be monitored by reference to specialist volumes. *General Trends in Shipping*, published as part of the *Business Monitor* series, supplies data on the UK merchant fleet, world fleets and international trade. *Basic Road Statistics* (British Road Federation), spanning over fifty years, contains annual information on most topics connected with roads and traffic. This account is especially valuable for the geographer as sections include regional and county statistics, as well as international comparisons. *The British Railways Board Annual Report and Accounts* affords comprehensive details on most aspects of rail transport. *UK Airlines* (Civil Aviation Authority) provides a monthly digest of operating traffic, and international passenger records are given in *Overseas Travel and Tourism* (*Business Monitor*).

Complementary to transport data are trade statistics. These provide descriptions of patterns of international trade, its composition and volume. Sources include *Overseas Trade Statistics of the United Kingdom* (Department of Trade and Industry), which shows total imports and exports by area and country of origin. *Port Statistics* (Department of Transport) has data from major port authorities on foreign and domestic traffic. This replaced the *Annual Digest of Port Statistics* which had been published from 1967–1980. A third publication, the *Statistical Abstract of United Kingdom Ports Industry* collates annual data from HM Customs and Excise. Additional trade data are available in the international digests of the UN, OECD and Eurostat.

Sources for the geography of economic growth and regional development

General appraisals of the state of the nation can be gained from a series of digests prepared by the Central Statistical Office. These can be used by geographers to provide essential background material for studies of regional problems and economic developments.

The Annual Abstract of Statistics for the United Kingdom, published annually since 1860, provides a valuable monitor of long-term trends and changes in the economy, demography and social conditions of the nation. In addition the *United Kingdom Balance of Payments* and the *United Kingdom National Accounts* present data for the last eleven years on how the nation makes and spends money, at home and overseas. For the most up-to-date statistics, reference can be made to the *Monthly Digest of Statistics*, *Financial Statistics*, *Economic Trends* and *Social Trends*.

Two sources which are especially valuable for observations of regional change and regional comparisons are *Regional Trends* and the *County Monitors and Reports* of the OPCS. Regional Trends is an annual publication dating back to 1965 which brings together detailed information highlighting regional variations in the UK. Data cover a wide range of social, demographic and economic topics presented by Standard Regions.

The County Monitors provide a breakdown of 1981 Census statistics for each county of England and Wales, each region in Scotland and for seven Special Areas designated as Inner Urban Areas in 1978 (i.e. Liverpool, Manchester-Salford, Birmingham, Newcastle-Gateshead, Lambeth, Hackney-Islington, London Docklands). For Northern Ireland only national coverage is available, although there is a special report for Belfast Local Government District. Each Monitor contains a summary of the final figures for population and housing and gives comparisons with previous Censuses. More detailed statistics are given in the full County Reports.

3.5
SECONDARY DATA SOURCES FOR SETTLEMENT GEOGRAPHY

Sources for the geography of housing

Information on the number of dwellings in different tenure categories (owner-occupation, private renting, local authority) and housing amenities are available in the *County Reports* and the *National Summary Volumes* (e.g. Great Britain: *Key Statistics for Urban Areas by County and Region*) of the 1981 Census. Since information on particular settlements is spread throughout a large number of volumes comparative analysis can be time-consuming. Two other sources allow quicker comparison of the housing characteristics of urban areas.

The first, *Housing and Construction Statistics of England and Wales*, published since 1972 by the Department of the Environment (D of E) provides the latest monthly and quarterly regional figures on aspects of construction, such as orders, investment and labour, and housing, including house building, reservations, slum clearance and stocks of dwellings. A second source, *Local Housing Statistics of England and Wales*, published jointly by the D of E and the Welsh Office as a quarterly journal since 1962, records data provided by individual local housing authorities and new town authorities. Other aggregations cover counties and Standard Regions. Tables include house building, housing improvement grants, council house sales and homeless households. Most of the data are expressed per 1000 householders, enabling quick comparisons of housing activity in local authorities of different size. Similar data are available in *Scottish Housing Statistics*.

The *National Housing and Dwelling Survey* of 1978 is a one-off survey which provides an additional insight into British housing in the late 1970s.

Sources for the study of the internal structure of urban areas

A broad range of sources provide information on the internal structure of towns and cities. Above all the *Small Area Statistics* of the Census offer the most comprehensive insight into contemporary intra-urban patterns. The two scales most appropriate for urban analysis are the enumeration district (ED), which is the smallest area, and the ward, which consists of a group of EDs. Each enumeration district consists of about 150 households or 500 people, whereas wards vary between 8000 to 30 000 people in size. Maps of the boundaries of EDs are produced on 1:10 000 OS maps. The map sheets cover an area of 5×5 km, that is, four sheets for each 10 km^2 National Grid Square. All 39 tables of the Census (additional tables for Wales and Scotland) are available at these levels of aggregation. Given the enormous data set the possibilities of different mapping exercises are very large. In this way the data in the Census can be used to test various models of urban structure. Other patterns may be revealed by examining the relationship between Census variables, such as distance from the city centre and levels of owner-occupation, or good/poor housing amenity and social class.

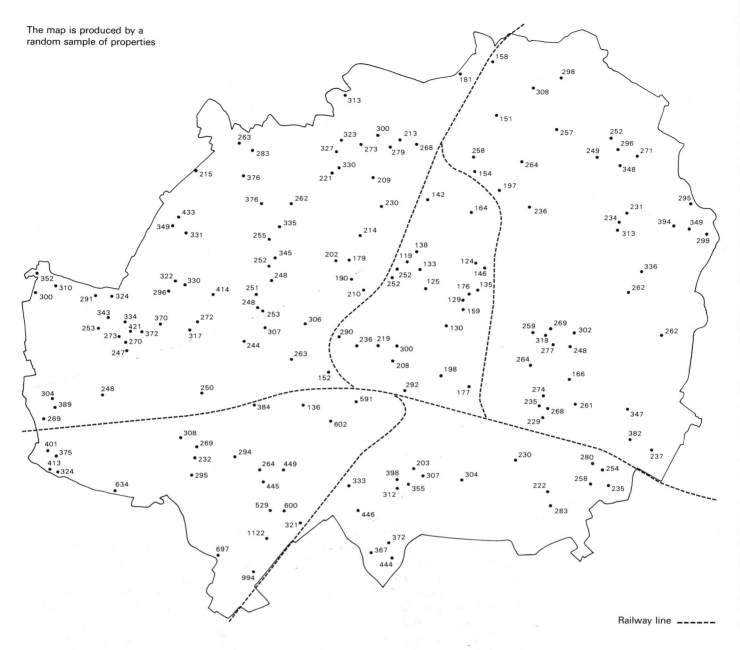

The map is produced by a random sample of properties

Railway line ------

Figure 3.10
Domestic rateable values (£) in Coventry

Local authorities are themselves useful sources of information about urban areas. The Town and Country Planning Act of 1947 required local authorities to prepare a plan of proposed land uses in the city for the next twenty years, subject to five-year revisions. These *development plans* contain maps and written statements of current and proposed land use. The advent of the 1968 Town and Country Planning Act required local *structure plans* to be produced and these further report on broad planning issues. These plans and their maps are very informative about the location of different types of land use. Both are usually available from the local planning department.

Furthermore, each local authority department may hold specialist information relating to their own activities. For example, the Highways Department may gather information on traffic flows and car ownership; the Housing Departments may collate details of housing improvement schemes. Before undertaking your project it may be worth checking with the local authority to see what data are publically available. Of course, the local authority may collect a great deal of information which is too sensitive or confidential to release.

Another local source are *rate books* which can be used both as contemporary guides and historical sources for studies of urban structure. The rates office of the local authority gathers information on the name and

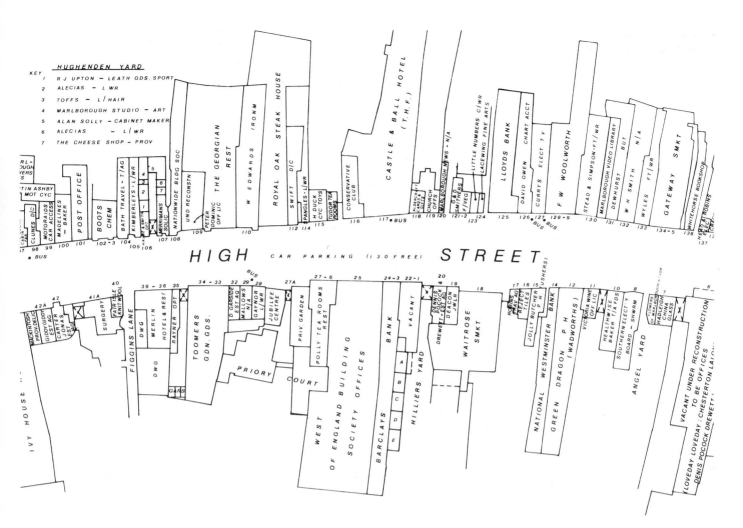

Figure 3.11
A Goad Plan of Marlborough, Wiltshire

address of a property, whether it is used for a commercial or non-commercial function and records its gross rateable value (GRV) and rateable value (RV). The GRV is derived by the valuation officer's assessment of the frontage, accessibility and physical condition of a property and the potential value of the site. The final sum is meant to represent a 'fair' annual rent which could be charged for the property and so is an indirect measure of land value. Rateable value is the figure which is used to calculate what the owners or occupants of the property must pay to the local authority. The differene between this amount and the GRV represents the likely annual cost of maintenance. The last assessment of GRV took place in 1973. The records are published for wards and are freely available for public inspection.

Rateable values can be used to provide an index of housing quality. In the absence of other data they are especially valuable for historical studies of the nineteenth century city, enabling areas of different status to be mapped. When using RV it is often more convenient to sample properties rather than attempt a complete inventory (Figure 3.10).

Functional patterns in a city can be established by careful scrutiny of rate books, more convenient sources are *Goad Plans* (Figure 3.11). Since 1967 Charles E. Goad (Salisbury Square, Old Hatfield, Hertfordshire AL9 5BE) have produced large scale plans of the shopping centres of all British towns with populations over 55 000. Some maps have also been produced of a number of suburban shopping areas within large settlements. These maps, drawn at a scale of 88 feet to an inch, provide information on street numbers, the area and frontage of properties, the name of the occupant and the type of retail activity. These maps are updated every four years and so can be used to look at change-over patterns. A limitation is that the maps only include information on the ground floor occupation of properties and so vertical variations of land use are missed.

Secondary data sources in physical geography

Fewer data in physical geography are available from officially published statistics. The *Guide to Official Statistics* (CSO, 1986) lists only 5 sections of relevance to physical geography: climate, pollution, forestry, water supply, and countryside and coastal areas. The Department of the Environment's *Digest of Environmental Protection and Water Statistics* (1977–) updates and synthesises some of the information on pollution of selected sites in the UK. Some statistics, on issues such as forestry and land use, are to be found in publications of Eurostat and the United Nations (especially the FAO, UNESCO and the World Meteorological Organisation). Information on the health effects of pollution are published by the World Health Organisation.

The following sections attempt to highlight sources of greatest use for project and classroom use.

3.6 SECONDARY DATA SOURCES FOR METEOROLOGY AND HYDROLOGY

Statistical sources for meteorology

Analysis of meteorological and hydrological information throughout the recent historical past has been made possible by the collection of some data since the late seventeenth century. Although earlier records were kept in Europe, the Royal Society of England encouraged observation of weather from the 1660s in observatories at Kew, Greenwich and Radcliffe. Many other long term rainfall records are available such as those (starting dates in brackets) at Wick (1850), Inverness (1841), Belfast (1819), Kendal (1829), Leeds (1824), Llandudno (1859), Manchester (1786), Mansfield (1807), Spalding (1726), Cambridge (1848), Oxford (1815), Kew (1697), Exeter (1817) and Falmouth (1835). By the middle of the last century, a more coordinated approach was adopted, and an annual publication, *British Rainfall*, was introduced. This journal is still available on an annual basis and lists approximately 600 of the 5000 stations for which the Meteorological Office collates and makes available complete weather records. Daily weather reports are available from the Meteorological Office covering some or all of the 54 stations given in Figure 3.13, but only for the period 1873–1980. Monthly weather reports have been published since 1884 and, at the present time, summary data are provided for 17 stations throughout the British Isles. For three of these stations, London (Kew), Manchester (Ringway) and Glasgow (Abbotsinch), graphs of daily temperature, sunshine hours and rainfall are given. The Meteorological Office should be approached directly for specific data, for which a charge is made (The Meteorological Office, London Road, Bracknell, Berkshire). Reference should also be made to the HMSO Sectional List No. 37 which lists Meteorological Office Publications. Official sources for climate statistics of the UK are shown below.

Climate Statistics: Meteorological Office Publications and other HMSO sources.

i *General*
Monthly Digest of Statistics (air temperature, rainfall, sunshine in UK)
Also see *Registrar General's Quarterly Return for N. Ireland*
       Annual Report of the Registrar General (N. Ireland)
       Annual Abstract of Statistics
       N. Ireland Annual Abstract of Statistics
       Scottish Abstract of Statistics
       Digest of Welsh Statistics

ii *Precipitation*
Monthly and Annual Totals of Rainfall for the United Kingdom
Also see *Snow Survey of Great Britain*
       The Meteorological Office Rainfall and Evaporation Calculation System (MORECS)

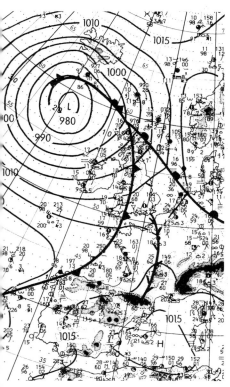

**Figure 3.12
A Meteosat photograph**

**Figure 3.13
Stations published in the Daily
Weather Report**

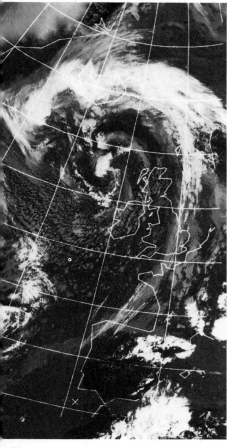

**Figure 3.14
A map published in the European
Meteorological Bulletin showing
the front in the Meteosat
photograph above**

iii *Marine Sources*

Marine Climatological Summaries (for sea areas and ocean weather ship
stations)

Since 1980, the *European Meteorological Bulletin* has been published
daily and provides information on general synoptic patterns for Europe
and, at a smaller scale, for the Northern Hemisphere. Individual stations
are not listed in the publication. The maps are available from Deutscher
Wetterdienst, Zentralamt, Frankfurter Strasse 135, Postfach 10 04 65,
6050 Offenbach A.M. FRG (Figure 3.14).

On occasions, individual researchers and, more recently specialist
organisations, have analysed and published meteorological data. One of
the most recent studies has been undertaken at the Climatic Research
Unit at the University of East Anglia, established by Professor H.H.
Lamb. They have compiled a historical archive of weather and climatic
statistics from the wealth of historical information available throughout
the world, including published statistics, weather diaries, ships logs and
newspaper reports. The result of one of the recent studies has been to
provide a generalised record of rainfall in England and Wales between
1766 and 1980 (Table 3.5 gives the figures for the period 1869–1980).

TABLE 3.5
Generalised record of rainfall for England and Wales

YEAR	JAN	FEB	MAR	APR	MAY	JUN	JUL	AUG	SEP	OCT	NOV	DEC	ANN.
1869	101.8	93.2	53.7	50.9	121.5	34.2	21.6	43.1	144.5	63.9	90.2	124.7	943.3
1870	65.9	62.1	55.8	21.5	40.0	31.4	39.4	60.3	49.8	159.3	66.8	80.8	733.1
1871	61.3	60.5	38.9	100.4	27.5	84.9	116.1	49.7	140.8	90.0	40.4	58.3	868.7
1872	140.2	88.2	84.9	69.1	66.2	100.0	122.9	80.8	104.1	149.5	140.7	138.3	1284.9
1873	109.1	51.3	75.8	23.1	61.6	62.3	84.4	98.8	72.0	97.7	71.7	26.6	834.5
1874	78.8	61.6	43.2	47.9	36.3	42.0	46.5	91.2	104.5	110.4	93.1	96.6	852.1
1875	122.8	46.5	27.2	39.1	52.0	92.8	142.7	67.5	96.2	157.3	145.6	47.5	1037.1
1876	47.1	97.7	99.4	84.7	23.6	55.6	47.2	83.5	141.5	74.4	108.4	193.9	1057.1
1877	146.8	77.6	76.3	93.6	79.5	46.3	108.5	136.1	74.3	86.0	141.5	77.6	1144.1
1878	68.4	50.6	38.7	69.2	117.6	76.4	39.3	155.2	72.6	104.4	118.0	73.1	983.6
1879	71.5	105.4	36.9	72.4	81.4	139.3	121.4	148.5	89.7	41.0	37.3	38.1	983.0
1880	16.2	84.1	52.1	56.4	38.6	81.3	157.5	40.1	119.8	161.4	97.9	122.8	1028.2
1881	35.5	109.9	81.9	34.2	53.1	68.9	75.3	143.7	73.0	82.0	105.9	103.3	966.5
1882	68.9	62.5	52.4	114.0	50.3	104.7	113.1	85.0	75.8	162.6	131.4	125.5	1146.2
1883	98.5	107.7	37.7	45.4	49.3	85.2	96.7	46.4	140.8	97.0	117.7	52.0	974.3
1884	108.7	70.7	66.4	47.2	39.1	41.3	97.5	47.7	66.4	51.8	51.2	103.5	791.5
1885	68.3	93.1	49.5	63.2	80.1	62.5	24.5	61.2	127.1	147.6	98.5	42.0	917.6
1886	120.2	29.5	77.6	59.0	123.1	31.9	105.3	52.2	72.1	135.6	99.4	145.2	1050.9
1887	71.8	24.1	49.8	36.6	54.0	20.8	38.1	53.4	94.2	71.4	86.1	69.0	669.3
1888	39.0	38.9	97.2	49.8	35.7	73.9	156.6	86.3	34.2	43.5	150.1	73.1	878.4
1889	38.2	60.2	78.0	80.3	94.8	23.2	87.8	98.3	58.8	127.9	43.5	58.2	849.2
1890	106.3	29.0	61.0	44.5	68.1	79.8	87.1	97.5	45.3	56.2	117.2	34.4	826.3
1891	67.2	3.6	59.0	45.0	83.2	53.9	84.6	149.3	65.6	162.3	90.7	132.2	996.6
1892	53.5	58.1	29.5	38.0	60.9	80.5	67.4	100.5	87.5	131.6	69.2	52.5	829.4
1893	57.8	103.5	15.8	9.9	45.8	38.9	104.4	61.4	58.2	98.6	74.0	87.9	756.3
1894	89.0	85.8	56.1	55.0	65.8	72.8	102.4	88.7	50.1	120.8	114.2	79.4	980.2
1895	100.9	11.1	73.9	55.2	17.7	38.1	116.9	92.4	23.3	113.3	122.2	90.8	855.8
1896	36.0	24.4	99.6	30.4	12.0	69.5	52.5	62.6	160.7	113.0	40.7	137.3	838.7
1897	61.7	90.8	108.8	69.3	39.8	77.7	36.5	117.5	93.9	38.2	69.2	112.4	915.8
1898	40.6	51.8	40.1	57.9	88.9	55.9	25.2	85.1	27.8	117.5	87.5	92.1	770.3
1899	119.3	75.5	35.6	79.8	70.1	43.2	49.8	39.9	92.9	81.6	79.2	73.2	839.9
1900	118.0	131.7	30.7	43.7	46.6	85.2	48.1	112.0	31.8	104.6	99.2	123.6	975.2
1901	56.3	41.7	70.4	73.8	35.0	57.2	71.9	68.0	56.0	72.3	66.5	136.1	805.2
1902	44.9	44.4	54.2	45.8	72.5	71.5	57.1	93.7	48.5	78.2	75.2	71.6	757.5
1903	96.7	58.8	109.5	58.5	83.7	82.9	103.9	122.7	102.9	218.1	70.1	72.5	1180.2
1904	99.0	112.1	50.7	46.1	74.4	32.8	74.9	82.4	58.1	44.4	62.7	77.6	815.1
1905	38.8	36.3	102.4	72.4	22.5	81.5	37.4	105.4	57.0	67.1	111.9	31.8	764.5
1906	135.7	78.7	57.6	29.4	84.0	66.2	34.0	70.1	33.7	146.9	109.8	80.1	926.3
1907	38.5	46.8	43.2	81.3	91.4	91.0	67.3	73.9	22.6	152.6	78.0	112.5	899.0
1908	55.7	54.6	91.9	76.2	60.1	36.1	85.4	94.8	69.0	56.2	48.6	78.2	806.7
1909	46.3	26.8	106.5	64.3	45.8	84.6	91.7	79.0	77.6	141.6	33.0	135.4	932.6
1910	90.6	103.4	27.2	68.9	63.9	70.0	83.7	106.4	16.4	100.0	128.0	138.5	997.0
1911	39.5	62.0	57.9	46.8	37.2	79.1	15.8	54.9	64.2	95.9	116.2	169.6	839.0
1912	107.2	60.9	118.4	9.0	56.6	122.4	94.4	192.9	50.0	98.7	76.3	111.8	1098.7
1913	124.4	34.0	106.8	98.2	68.6	39.9	32.6	41.9	65.6	105.5	96.7	54.2	868.4
1914	48.6	82.1	119.7	37.0	47.9	61.8	94.2	61.5	48.5	68.9	109.5	190.8	970.5
1915	108.7	123.5	35.4	35.7	69.8	35.0	126.0	76.7	47.0	72.7	75.0	184.9	990.4
1916	56.6	132.0	107.1	47.4	71.7	70.4	55.4	88.8	51.9	138.9	111.9	95.9	1027.7
1917	58.4	31.9	68.4	54.3	60.7	65.8	65.1	172.0	55.5	129.6	60.3	41.1	863.1
1918	91.2	61.6	36.7	63.3	58.5	30.9	110.0	62.9	189.5	69.1	66.3	122.5	962.6
1919	122.4	79.0	120.0	66.3	25.4	36.5	57.9	79.0	52.4	62.3	77.8	146.7	925.7
1920	108.3	43.0	82.0	116.5	74.6	66.9	132.4	50.6	71.1	70.9	44.9	92.1	953.4
1921	107.2	10.2	58.5	33.3	46.5	10.3	29.3	93.5	40.9	52.2	70.9	76.2	629.1
1922	109.4	89.7	68.0	82.3	34.0	40.4	126.3	100.3	79.4	33.4	48.7	116.4	928.4
1923	63.5	152.7	52.8	70.6	69.8	19.4	79.4	92.7	82.5	137.4	92.3	100.0	1013.1
1924	94.1	27.6	37.6	72.9	122.4	67.2	118.4	98.5	114.6	132.7	72.8	123.9	1082.7
1925	82.3	124.2	32.8	74.2	96.5	4.3	81.3	91.0	98.6	106.7	76.6	98.8	967.3
1926	114.1	80.3	24.5	65.2	69.3	81.1	76.2	66.0	58.1	87.6	157.9	24.5	904.9
1927	94.7	80.5	89.0	59.2	37.8	101.3	96.0	139.5	154.0	70.8	101.9	83.5	1108.1
1928	160.3	76.3	78.6	39.2	40.0	97.0	60.3	95.2	29.8	144.4	112.1	90.0	1023.0
1929	49.1	31.5	8.0	34.9	62.6	46.7	66.4	71.8	23.6	124.3	195.9	179.0	893.8
1930	125.1	26.1	65.3	71.0	62.0	33.1	109.1	104.4	110.8	92.4	128.1	97.5	1024.8
1931	79.9	80.0	27.6	93.2	93.8	90.8	109.1	119.2	76.9	32.3	132.1	45.4	980.2
1932	87.8	9.2	59.3	83.9	125.0	34.6	104.0	49.3	94.3	161.9	63.1	47.4	920.0
1933	65.8	94.6	75.0	37.9	59.7	56.3	62.4	32.4	56.1	103.1	47.3	29.0	719.4
1934	76.0	12.4	71.9	73.1	39.5	45.8	49.2	81.1	69.9	82.8	64.2	182.8	848.8
1935	43.6	101.0	26.2	103.6	33.4	91.6	28.5	64.4	138.0	132.6	150.8	94.4	1008.2
1936	124.9	75.4	58.0	55.2	31.8	97.0	139.3	30.9	93.6	62.2	106.6	88.9	964.0
1937	128.8	138.9	92.0	80.4	78.4	42.3	63.9	40.4	64.1	83.5	49.4	97.5	959.5
1938	110.1	33.9	17.0	6.5	72.9	57.7	82.9	89.8	58.8	131.7	120.2	108.6	890.1
1939	158.5	54.2	59.1	67.4	33.1	62.1	121.6	73.4	31.8	122.7	158.8	61.8	1004.6
1940	73.3	78.0	80.3	60.9	39.2	21.8	113.8	16.5	48.5	108.1	196.4	74.8	911.7
1941	93.3	100.7	90.8	38.8	65.5	32.9	76.0	124.3	20.0	75.7	81.6	56.1	855.7
1942	106.7	29.7	63.6	42.5	103.4	20.7	75.9	84.5	69.1	96.5	40.5	99.4	832.4
1943	147.3	47.0	26.9	37.4	84.3	60.3	55.3	83.4	85.8	83.2	68.8	51:8	831.5
1944	82.5	38.9	11.9	56.5	38.1	64.5	70.6	79.1	106.1	113.9	154.0	69.7	885.7
1945	84.0	78.9	31.6	44.7	81.3	81.4	72.5	68.4	64.8	103.2	18.7	104.2	833.7
1946	85.0	84.7	38.2	46.1	79.3	83.2	75.9	141.5	126.6	39.7	157.1	93.4	1050.7
1947	82.2	56.3	174.3	74.9	59.2	68.4	67.5	12.8	47.4	22.7	81.6	80.7	828.1
1948	174.6	48.2	34.0	54.5	71.1	93.1	44.3	120.7	71.0	77.1	52.8	108.2	949.6
1949	41.7	40.8	43.8	61.6	65.4	19.5	54.0	51.0	48.4	157.8	108.9	86.7	779.7
1950	40.5	140.8	39.1	70.1	54.3	46.1	107.8	123.3	123.1	40.7	151.5	71.2	1008.6
1951	97.7	111.1	113.1	71.6	78.3	34.7	50.6	126.8	81.1	34.3	174.4	94.6	1068.4
1952	84.0	25.5	73.5	61.6	65.3	53.8	36.7	104.5	91.9	104.3	98.4	92.2	891.7
1953	31.8	50.5	29.1	70.7	62.7	62.4	98.4	87.9	80.7	69.3	68.4	35.3	747.1
1954	62.2	82.2	77.6	16.4	81.6	92.1	89.7	125.5	97.0	118.5	163.4	93.6	1099.9
1955	85.9	64.1	52.5	32.5	102.9	86.8	28.9	25.7	53.0	72.3	57.4	115.8	778.0
1956	122.0	26.0	30.7	41.4	22.4	71.7	104.5	159.8	95.7	54.1	34.4	108.8	871.5
1957	73.9	103.6	72.3	9.2	46.4	53.6	100.0	107.5	123.8	74.5	63.1	78.1	906.0
1958	94.9	116.6	50.5	31.5	83.7	108.3	97.9	103.4	120.2	81.2	58.6	105.2	1052.0
1959	103.2	9.1	70.9	77.6	26.4	43.3	70.2	35.5	8.4	91.1	127.3	163.1	826.2
1960	125.0	80.3	52.7	44.8	46.6	50.9	120.2	106.4	114.2	177.9	153.0	118.3	1190.4
1961	117.9	72.1	14.4	102.8	38.2	39.4	68.0	85.6	78.0	116.4	62.7	103.2	898.8
1962	113.2	35.8	47.5	71.2	67.4	17.4	60.6	107.4	106.4	37.3	69.5	75.2	808.7
1963	31.0	35.3	101.9	80.8	48.2	85.7	55.5	115.6	72.1	62.0	170.9	29.2	888.2
1964	27.6	34.3	100.9	68.3	58.6	79.8	50.9	54.3	34.6	64.6	59.5	98.5	731.9
1965	99.5	16.4	74.7	61.0	65.9	71.5	98.8	74.7	138.5	32.9	117.4	170.9	1022.2
1966	60.4	127.0	34.6	111.2	68.6	88.3	74.6	109.4	54.0	129.1	77.5	114.9	1049.6
1967	62.5	90.6	54.5	46.2	138.9	40.0	64.0	75.1	100.3	170.7	73.5	77.9	994.4
1968	86.9	47.6	58.2	70.2	78.5	91.6	102.8	73.6	151.5	96.9	71.8	85.8	1015.4
1969	106.5	69.0	79.9	58.7	121.5	53.6	69.9	75.1	41.9	17.4	139.6	88.9	922.0
1970	111.5	92.8	66.3	87.7	25.1	43.4	75.2	80.9	67.3	60.7	183.7	54.2	948.8
1971	110.5	36.8	71.0	52.8	50.4	105.2	45.8	115.0	27.2	77.6	100.0	40.2	832.7
1972	106.5	80.3	78.1	70.1	75.2	74.4	53.1	36.4	41.2	36.0	109.0	110.5	870.9
1973	47.8	46.7	23.1	67.4	83.8	60.9	92.8	65.0	76.9	56.8	53.1	69.1	743.3
1974	122.7	107.3	48.0	14.9	41.2	68.2	75.2	95.5	154.5	93.3	131.9	78.2	1030.9
1975	120.1	32.9	82.2	69.6	49.9	21.0	61.6	53.2	107.4	38.2	74.0	48.3	758.4
1976	58.5	39.8	49.6	19.7	64.0	17.5	29.8	26.5	150.3	160.1	88.0	96.9	800.8
1977	104.4	144.2	74.8	51.2	51.0	86.4	24.6	101.5	41.6	71.4	107.3	96.0	954.3
1978	110.2	89.6	86.2	54.3	46.6	59.7	84.3	65.9	54.5	19.6	56.4	171.7	898.9
1979	89.8	70.0	128.8	70.5	120.4	40.1	34.3	93.3	39.1	87.8	90.3	161.0	1025.4
1980	77.2	93.7	103.6	17.7	32.5	127.1	73.0	92.3	65.4	131.3	92.7	77.6	984.1

These data will be used in a later section of this book as a basis for statistical exercises (Chapter 5).

Statistical sources for hydrology

Hydrological data are less well represented in either map or statistical form at national and international level. The Geological Survey published some of the earliest information on borehole and well level data. These data were collated, along with meteorological records, in the *County Water Supply Memoirs*, first published in 1899. The *Groundwater Year Book* (HMSO, 1964–1973) and *National Riverflow* (1974–1980) provide summary statistics in relation to the borehole observation network.

Until the formation of the river boards in 1948 (Table 3.6) there was no statutory requirement to collect river flow data. The longest records come from Teddington Weir on the River Thames which has been gauged continuously since 1883. The stations for which records are available are listed in the *Surface Water Yearbooks* (HMSO, 1964–1973) and in *National Riverflow* (1974–1980), both collated by the Water Data Unit until its recent closure.

TABLE 3.6
River management organisations (after Potter, 1978)

Organisation and operative dates	Regional coverage and area of responsibility	Reporting policy
Water authorities 1974–	10 to cover England and Wales water supply, river management and river and effluent quality	Very little technical detail in annual report. Flood events are often subject of a separate report
River authorities 1965–1974	29 to cover England and Wales. Water resource planning, river management and river quality	Floods and other hydrological and meteorological reports. Some separate reports for flood events. Internal committee dealing with land drainage and water resources
River boards 1951–1965	32 to cover England and Wales. Primarily river management with some water resources responsibility	Reporting policy similar to river authorities
Catchment boards 1931–1951	46 to cover about 67% of England and Wales. River management and land drainage	Annual or triennial reports produced spasmodically by some and include mainly flood and flood protection information
Fishery boards reconstituted 1932	Cover major fishing rivers in UK	Annual reports produced and include river level, rainfall and temperature and pollution information
Internal drainage boards 1861–	Currently 314 to cover 12500 km^2 of low-lying areas adjacent to larger rivers and fen districts	Only the largest authorities produced annual reports, and these concentrate on floods and flood protection measures. Engineer of authority would report internally on notable flood events
River conservancies (1857–1930)	A few catchments including Thames, Lea, Cumbrian, Derwent and Dee. Maintenance of major river channels	Occasional reports on particular floods and annual reports
Sewer commissioners (various acts from 1427)	Responsible for specific river reaches, sponsoring improvment works and involved in cases of dispute	

As a result of the closure of the Water Data Unit, the groundwater and surface water data have been transferred to the British Geological Survey (BGS) and the Institute of Hydrology (IOH) respectively. Both are located at Crowmarsh Gifford, Wallingford, Oxon OX10 8BB. Since 1981, groundwater and river flow statistics are available in *Hydrological Data UK*, available from the IOH. (Water pollution data are available from alternative sources.) Some review data on international rivers are published by UNESCO, and for the UK the HMSO publishes statistics on water use and demand.

River discharge data: UNESCO

Discharge of selected rivers of the world

Vol I   General and regime characteristics of selected stations
Vol II Monthly and annual discharges recorded at various selected stations (up to 1964)
Vol III Pt 1 Mean monthly and extreme discharges (1965–69)
Vol III Pt 2 Mean monthly and extreme discharges (1969–72)

List of International Hydrological Decade Stations of the World

UK water supply statistics:

Annual Report of the National Water Council
Municipal Year Book
*Water Authority Annual Reports**
Waterfacts

*Many of the regional Water Authorities have an information officer who may be able to provide you with useful data for your local area or problem.

47

Graphical sources for hydrology and meteorology

A further source of hydrological and meteorological information is the *Soil Survey*. In England and Wales the Soil Survey publishes small scale (1:1 000 000) maps of:

   i   median accumulated temperature above 0°C Jan–Jun;
  ii mean maximum potential soil moisture deficit;
 iii average maximum potential cumulative soil moisture deficit;
 iv winter rain acceptance potential;
  v wind exposure;
 vi median duration of field capacity;
vii bioclimate.

Three climatically based maps are also available from the *Soil Survey of Scotland* at the 1:625 000 scale:

   i bioclimate;
  ii exposure and accumulated frost;
 iii temperature and potential water deficit.

It is important to remember that the maps obtained from the Soil Survey organisations are produced with a strong bias towards conditions which limit crop growth.

Media sources for hydrology and meteorology

Newspapers are particularly valuable as a historical source of information for physical geography. Many studies using newspapers have revealed important historical information on meteorological phenomena and natural hazards, such as river and coastal floods and storms, earthquakes, landslides and droughts. Newspapers date back to the early eighteenth

**Figure 3.15
Provincial evening newspapers in Britain**

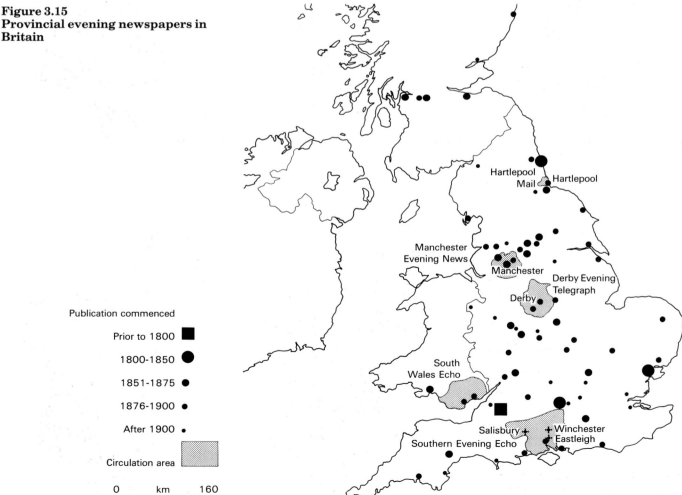

Publication commenced

Prior to 1800 ■

1800-1850 ●

1851-1875 ●

1876-1900 ·

After 1900 ·

Circulation area

0   km   160

Hartlepool Mail — Hartlepool

Manchester Evening News — Manchester

Derby Evening Telegraph — Derby

South Wales Echo

Salisbury — Winchester — Eastleigh

Southern Evening Echo

Mon Jan 7th 1740: 'The river of Forth is now so frozen up beyond what has been known in the memory of man.'
Mon Jan 21st 1740: 'On Friday a sailor was found frozen to death on board a sloop in Leith harbour. Then followed an entertaining story of a lad who slipped and fell while carrying a 20 pint cask of small beer. The cask was dashed to pieces, but the liquor was so completely frozen that like a nut turned out of its shell, he tied a rope about it and brought it to town'.

century, but it is only in the last 130 years, from the repeal of Stamp Duty on publishers in 1855, that a large number of local newspapers emerged. The 73 regional evening newspapers analysed by Gregory and Williams (1981) provide over 5000 years of information (Figure 3.15).

Contemporary data can be collected from local and national newspapers. Some of these sources provide detailed meteorological information and even produce a weather map which may be a valuable supplement to data derived from satellite images, such as *Meteosat* (Figure 3.12) or from BBC or independent radio and television broadcasts.

Newspaper reports can be used to obtain a range of different types of information. This will include individual event-based information such as the dramatic reports of Scottish snowstorms given in the margin (left). These extracts are taken from a report by M.G. Pearson of the 1739–40 winter in Scotland published in *Weather* (1973). Original extracts were taken from the *Caledonian Mercury*.

Newspaper records may also be used to examine long term trends in particular hazards, such as the record of flooding in Derbyshire based on the number of reports given by the *Derby Evening Telegraph* (Figure 3.16).

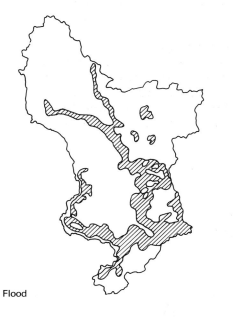

Flood

Gales
thunder

Drought
landslip
earth tremor

0   km   30

Hazard zones

Flood

Gales

Thunder

Drought

Landslip

Earth tremor

Snow

Frost/ice

Fog

Buxton
Chesterfield
Matlock
Ashbourne   Heanor
Derby

Snow

Frost/ice
fog

Figure 3.16
Hazard distribution in Derbyshire based upon reports in the *Derby Evening Telegraph*, 1879–1978

Figure 3.17
Coastal stations and sea areas used in BBC shipping forecasts with (opposite) a Weather Forecast Chart

South-East Iceland

Faeroes

Fair Isle

Viking

North Utsire

South Utsire

Fisher

Su

German Bight

Forties

Crom-arty

Bailey

Hebrides

BL

Dogger

Forth

StA

Tyne

Humber

D

Thames

T

Malin

M

R

Irish Sea

Do Dover

Rockall

Wight

Lundy

Port-land

C J

Va

Fastnet

LE

Plymouth

Shannon

Sole

Biscay

Finisterre

Stations whose latest reports are broadcast in the 5 minute forecasts

T	Tiree
BL	Butt of Lewis
Su	Sumburgh
StA	St Abb's Head
D	Dowsing
Do	Dover
C	Channel Light-vessel
LE	Land's End
Va	Valentia
R	Ronaldsway
M	Malin Head
J	Jersey

BBC RADIO 4 WEATHER FORECAST CHART

GENERAL SYNOPSIS at _____ GMT/BST

System	Present position	Movement	Forecast position	at

SEA AREA FORECAST

Gales	SEA AREA FORECAST		Wind (At first)	Wind (Later)	Weather	Visibility
	VIKING	1				
	N UTSIRE	2				
	S UTSIRE	3				
	FORTIES	4				
	CROMARTY	5				
	FORTH	6				
	TYNE	7				
	DOGGER	8				
	FISHER	9				
	GERMAN BIGHT	10				
	HUMBER	11				
	THAMES	12				
	DOVER	13				
	WIGHT	14				
	PORTLAND	15				
	PLYMOUTH	16				
	BISCAY	17				
	FINISTERRE	18				
	SOLE	19				
	LUNDY	20				
	FASTNET	21				
	IRISH SEA	22				
	SHANNON	23				
	ROCKALL	24				
	MALIN	25				
	HEBRIDES	26				
	BAILEY	27				
	FAIR ISLE	28				
	FAEROES	29				
	SE ICELAND	30				

COASTAL REPORTS at _____ GMT/BST

COASTAL REPORTS	Wind Direction	Force	Weather	Visibility	Pressure	Change
T Tiree						
BL Butt of Lewis						
Su Sumburgh						
StA St Abbs Head						
D Dowsing						
Do Dover						

COASTAL REPORTS	Wind Direction	Force	Weather	Visibility	Pressure	Change
C Channel Light-vessel						
LE Land's End						
Va Valentia						
R Ronaldsway						
M Malin Head						
J Jersey						

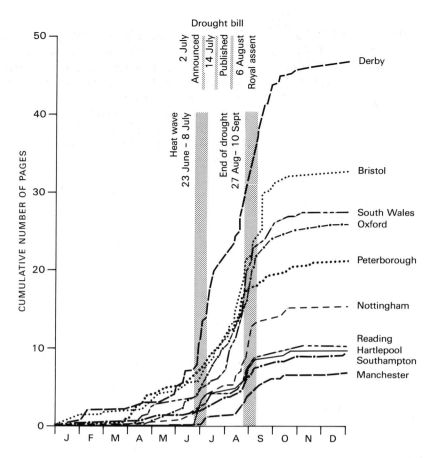

**Figure 3.18
Cumulative frequency of references
to drought in ten evening
newspapers 1976**

Spatial patterns of hazard impact can also be established on a county or national basis for more recent events (Figure 3.18).

If newspapers are used to collect secondary information, it is important that you know the kind of information you are looking for and, where possible, the dates on which events occurred. In particular, you are advised to search two important local sources. First, many newspapers and journals are available on microfilm and are often to be found in study rooms of major city and county libraries and/or in college and university libraries. Secondly, if a local newspaper is still in circulation, you are advised to consult back issues maintained in the office's archives. Your local library may also be able to help you track down any published indexes for national newspapers although these are rare; some indexes are kept in the British Museum.

Often neglected sources of secondary meteorological data are the broadcasting media. Television and routine broadcasts are often ignored because high quality statistical information is also available directly from the Meteorological Office. However, a worthwhile study would be to 'calibrate' newspaper and broadcast information against local weather reports, as a means of checking the validity of the information and especially the forecasts obtained. Since the discontinuation of *Daily Weather Reports* in 1980, newspapers and, in particular, the coastal and shipping forecasts broadcast on the BBC long wave may be used to compile an accurate daily chart. Figure 3.17, which shows coastal stations and sea areas and the record table, may be used in conjunction with a recording of the broadcast to compile daily meteorological events. This information will enable you to evaluate the reliability and accuracy of forecasts for your own area. More detailed information and local weather forecasts are available on request from coastguard stations, airports and local meteorological offices. *Ceefax* and *Oracle* teletext services provide general updates and some of you may also have access to satellite images received from *Meteosat*.

3.7
SECONDARY DATA SOURCES FOR ENVIRONMENTAL POLLUTION

Environmental pollution data sources may be found in the *Digest of Environmental Statistics* (D of E). This provides an annual report summarising trends in air pollution, water quality, radioactivity, noise, blood lead levels, solid waste disposal, water supply and use, landscape and nature conservation.

These data are updated annually and provide important reviews of historical trends (Figure 3.19). Other sources of general information include *Regional Trends* and the *Royal Commission on Environmental Pollution*.

Figure 3.19
Two examples of data collected by the Department of the Environment

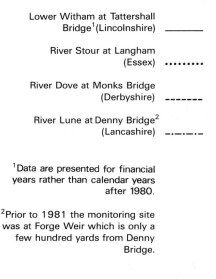

Lower Witham at Tattershall Bridge[1] (Lincolnshire) ⎯⎯⎯

River Stour at Langham (Essex) ·········

River Dove at Monks Bridge (Derbyshire) ⎯ ⎯ ⎯

River Lune at Denny Bridge[2] (Lancashire) ⎯·⎯·⎯·

[1]Data are presented for financial years rather than calendar years after 1980.

[2]Prior to 1981 the monitoring site was at Forge Weir which is only a few hundred yards from Denny Bridge.

Emission from coal combustion

Domestic ▨

Industry (including railways) ▨

Average urban concentration of smoke from all sources ⎯⎯⎯

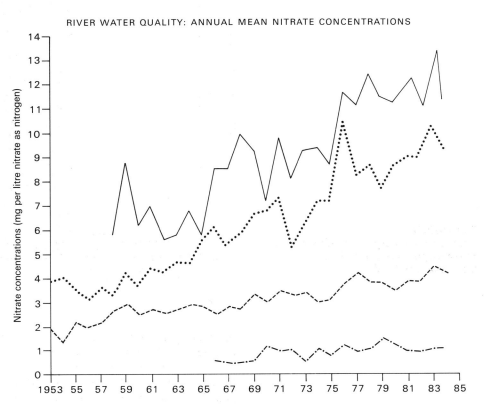

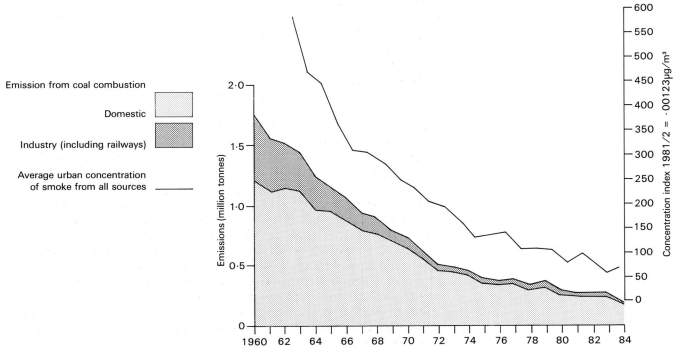

Air pollution sources

These may be derived from HMSO publications such as *National Survey* (smoke and sulphur dioxide), *Health and Safety* (industrial air pollution), *Annual Report of the Chief Alkali Inspector* (N. Ireland), *Transport Statistics, Great Britain*.

Much of the data have been collated since 1914 by the Department of Industry Laboratory (Warren Spring Laboratory, PO Box 20, Gunnels Wood Road, Stevenage, Herts, SG1 2BX). Since 1981, this laboratory has ceased collating and publishing smoke, sulphur dioxide and dry deposition data. For your own local area, these statistics may be available from the Environmental Health Office.

Water pollution sources

For the freshwater environment, some regular statistics are published annually by Water Authorities in their annual reports. National data from the Harmonised Monitoring Network are available from the HMSO at Romney House, 43 Marsham St., London SW1P 3PV. Some special publications and occasional reports may give useful data, such as the *Freshwater Survey of England and Wales* (updated 1975), the *Digest of Welsh Statistics* and the *Port of London Authority Annual Report*.

In the marine environment, the *Aquatic Environment Monitoring Report* and the *Freshwater Pollution Survey of England and Wales, Volume 3* (with coastal pollution) are particularly important. *Waterfacts* contains a classification for estuaries, as well as some freshwater data.

Sources for other forms of pollution

Land pollution and noise data are available from the *Digest of Environmental Protection and Water Statistics* (HMSO). Specific information with regard to radioactivity may be found in sources such as *Environmental Radioactivity Surveillance Programme* (NRPB; annual results for the UK), *Radioactivity in Surface and Coastal Waters of the British Isles* (HMSO), *Annual Report on Radioactive Discharges and Monitoring of the Environment* (HMSO).

3.8 SECONDARY DATA SOURCES FOR GEOMORPHOLOGY

Secondary data sources are less widely available for the geomorphologist than the hydrologist and meteorologist since few national data collection agencies exist. In general, two major sources are relevant. These are *map sources*, especially topographic and geological maps, and Admiralty charts, and *aerial photographs*, which have been available for many areas since 1945. Some additional photographic and documentary records may be available but in general this needs to be established locally through such agencies as County Record Offices and county and city reference libraries.

Map sources for geomorphology

A review of the availability and scales of maps published by the Ordnance Survey and other agencies has already been given (section 3.1). *The Geological Survey of Great Britain*, now known as the *British Geological Survey* (BGS), began its mapping operation in 1835 and has covered most of Great Britain and Northern Ireland at the 1:63 360 scale. These maps are currently being revised, in line with the Ordnance Survey change to the 1:50 000 scale, although few areas are as yet covered at this scale. The 1:63 360 and 1:50 000 scale maps are accompanied by sheet memoirs and explanatory booklets. In addition to these large scale maps, the BGS publishes a range of smaller scale maps at between 1:625 000 and 1:250 000. These include geothermal maps, generalised solid geology, quaternary geology and hydrogeological maps of the UK. Between these small scale maps and the 1:50 000 scale maps are a range of regional geology maps of solid and quaternary deposits. Some particularly interesting areas are mapped at the 1:250 000 scale and are associated with explanatory booklets. Further information on the availability of geological maps can be obtained from The Information Office,

TABLE 3.7
Aerial photograph scales
and coverage

Scale	Coverage Sq kms
1:7500	3.0
1:10 000	5.3
1:20 000	21.0
1:60 000	190.0

Some larger scale photographs may be available of
urban areas, but not for rural areas.
Enlargements up to 4× may be done
photographically, increasing the scale of the
1:75 000 to better than 1:2 000. At this scale,
features of 210m in length are 1cm long on the
photograph. Historical sequences in active
environments, such as coastal spits and sand bars,
or regions of intense river erosion can be mapped
with great accuracy from photographs enlarged to
this scale.

Wales:
Air Photographs Unit,
Welsh Office,
Room G-003,
Crown Offices,
Cathays Park,
Cardiff.

N. Ireland:
Ordnance Survey of Northern Ireland,
Department of the Environment,
83 Ladas Drive,
Belfast,
BT6 9FJ.

Scotland:
Air Photographs Unit,
Scottish Development Department,
Room 1/24,
New St Andrew's House,
Edinburgh,
EH1 3SZ.

England:
Ordnance Survey,
Air Photo Cover Group,
Air Survey Branch N152,
Romsey Road,
Maybush,
Southampton,
SO9 4DH.

British Geological Survey, Geological Museum, Exhibition Road, London SW7 2DE.

For studies in coastal geomorphology, additional and very important sources of information are the *Hydrographic Surveys* undertaken by the Admiralty. Marine charts are to be found for certain areas since the sixteenth century. However, widespread coverage was not available until the early eighteenth century following the establishment of the Hydrographic Department of the Admiralty in 1795. Throughout the nineteenth century, harbour and coastal charts were published at scales ranging from 1:63 360 to 1:3168. Lists of contemporary Admiralty charts are available from the Ordnance Survey.

Aerial photography
Particularly useful sources of geomorphological data and land use information since 1945 are air photographs. In general, photographs are available at four scales printed on a 23 × 23 cm frame shown in Table 3.7. Since 1975, registers of air photograph cover have been compiled from RAF, Ordnance Survey and commercially flown coverage. These are available from the addresses given in the margin. If you are trying to get information for a specific area, you will need to send either exact grid references or a tracing of the 1:50 000 map showing what information and aerial coverage you want.

An appraisal of map and photographic sources of information
Maps and photographs have to be sampled in order to obtain the desired information. Care should be taken to select both map scale and type of map or photograph. This can be illustrated with reference to some exercises you may wish to carry out from this secondary data source.

Comparison of features on maps should be done at the same scale because different levels of detail are available at different scales. Even at the same scale, different editions of the map may provide you with different information. As an example, examine the river network given in Figure 3.20. Figures a and b are taken from 1:63 360 scale maps published in 1809 and 1961. The networks have been numbered by a standard method of *Stream Ordering* which is often useful for studying river networks. It is done by assigning the value 1 (*first order*) to unbranched streams. Where two first order streams join, we have a second order stream and so on. (If you use this method remember that you can only increase the order of the stream when two streams of the SAME ORDER join and you only count a stream of a particular order ONCE). The results of this analysis are summarised in Table 3.8.

Because of these variations, it is often difficult to examine changes in river networks and drainage basin properties at the 1:50 000 or 1:63 360 scale and although these scales are useful for looking at more general landscape changes, most geomorphologists collect data from maps published at the 1:25 000 scale or larger where more detail is available. A further complication with pre-First World War Ordnance Survey maps is that, as explained earlier, different map projections were used. An example of such a map is given in Figure 3.2.

The 1:25 000 scale is probably the most useful and allows us to collect information not only about basin networks, but about a range of properties of the drainage basin. Some examples are listed in Table 3.9 along with data on factors such as valley shape and asymmetry which can be examined by drawing scale profiles taken from the map.

More detailed studies of historical changes can only be performed with larger scale maps. Figure 3.21, for example, shows how to study the changes in a river course for more than 150 years with data taken from Tithe maps and Ordnance Survey maps of an area made at a scale of 1:2500. Figure 3.22 shows how to make use of Admiralty hydrographic charts for the reconstruction of important changes which may be found in coastal areas around the British Isles.

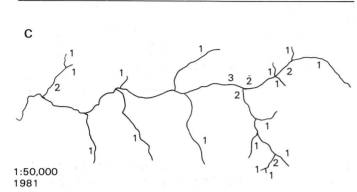

a

1

Old series OS maps. 1″
surveyed-mostly from 1809

b

1:63360
1961

c

1:50,000
1981

d

1:25,000
1959 survey

Figure 3.20
A river network shown on different
scales and at different periods of
time

TABLE 3.8
Analysis of stream networks

Scale of map	Date of publication	No. of streams of Order		
		3	2	1
1:25 000	1959	1	3	24
1:50 000	1981	1	3	16
1:63 360	1961	1	3	13
1:63 360	1809	1	4	15

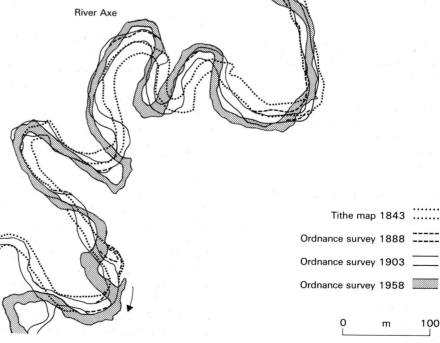

Figure 3.21
The changing course of the River
Axe revealed on large scale maps
over a period of time

River Axe

Tithe map 1843 ·······
Ordnance survey 1888 - - - -
Ordnance survey 1903 ———
Ordnance survey 1958 ▨

0   m   100

TABLE 3.9
Some easily measured properties of
the drainage basin

Property	Units	Comments
Basin area	km^2	Trace the basin outline from map
Stream length	km	Trace the stream network from map
Drainage density	$km\,km^{-2}$	Divide stream length by basin area

* These properties can also be measured for the
basins ordered by the method explained above.

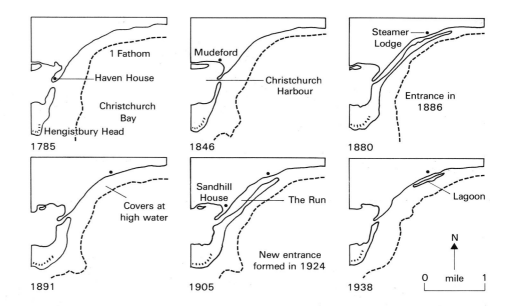

Figure 3.22
Admiralty hydrographic charts have revealed the changing shape of the entrance to Christchurch Harbour

3.9
SECONDARY DATA SOURCES FOR BIOGEOGRAPHY

Biogeographical information is largely restricted to details of land use and soils published in map form such as the Land Utilisation Surveys of the 1940s and 1960s described earlier. Some information may also be obtained from aerial photographs, although interpretation is difficult because most photographs are in monochrome. Additional material can be obtained from remotely sensed satellite images (Figure 3.23) and from specialist organisations such as the Forestry Commission and the Ministry of Agriculture, Fisheries and Food (MAFF).

Land use and soil
Early land use information can be obtained from estate maps and the Tithe surveys. Availability of estate maps will have to be traced in County Record Offices. In 1966, it was estimated that at least 20 000 of such maps were in existence, but few catalogues are as yet available. In many instances, several days of research will prove less than fruitful and you are advised to try to follow this route with great caution. Tithe maps were usually drawn at a large scale (1:2376 or 3 chains to an inch). Historical land use information of a general nature can be found in the *Survey Books of the Ordnance Survey* (1870–). Detailed agricultural data can be derived from a study of the returns of the Agricultural Census, available to the public at a parish level.

Information relating to soils is currently incomplete at a large scale. The Soil Survey of England and Wales and the Soil Survey of Scotland have compiled and published a variety of maps, usually in association with regional bulletins, following the model provided by the Geological Survey. County and District surveys are published at either 1:63 360 or 1:50 000 scales and some areas are covered at the 1:2500 scale. For England, Wales and Scotland, all areas are mapped at the 1:250 000 scale (a total of 13 maps) which may form the basis of many useful studies. In addition land use capability maps are published by both the Soil Survey of England and Wales by the Soil Survey of Scotland at a scale of 1:25 000. These maps classify land capability into seven groups depending on management limitations such as stoniness, permeability, soil depth, gradient and wetness. Agricultural land classification maps are published by MAFF at a scale of 1:63 360. The five categories in the classification are based upon similar characteristics to those used by the Soil Survey. These include climate, especially rainfall and temperature, topography and gradient, and a range of soil properties such as moisture deficit and texture. Further details may be obtained from these organisations, and their addresses are given in the margin.

The Soil Survey of England and Wales,
Rothamsted Experimental Station,
Harpendon,
Hertfordshire.

Department of Soil Survey,
Macaulay Institute for Soil Research,
Craigiebuckler
Aberdeen,
AB9 2QJ.

Other maps and reference material may be obtained from:

MAFF,
Lion House,
Willowburn Estate,
Alnwick,
Northumberland.

The Welsh Office,
Map Library,
Ground Floor,
Crown Building,
Cathays Park,
Cardiff,
CF1 3NQ.

Forestry records

Records of forestry are available at a national and international scale. Summary statistics for the UK are contained in the *Annual Report and Accounts of the Forestry Commission* published since 1920. This report contains a general review and detailed statistics on forestry enterprise (planting, harvesting, conservation) for England, Scotland and Wales, further sub-divided on a regional basis. Some European policy strategies and reports are also included, along with addresses of national, regional and local offices.

Other statistical sources include: *Forestry in Northern Ireland; Monthly Digest of Statistics* (total area of woodland on agricultural holdings); *Scottish Abstract of Statistics; Digest of Welsh Statistics; Census of Woodlands 1979–82* (information about Britain's forest resources, type and ownership, on a county basis in England and Wales and on a regional basis for Scotland).

International forestry statistics are published by the UN (FAO) in its *Annual Digest of Statistics*. More recently, the International Institute for Environment and Development have made an assessment of the resource base that supports the global economy published as *World Resources* (1986). It includes data on erosion and agriculture, forests and vegetation, wildlife and habitat and provides data tables for 146 countries.

Other information relevant to biogeographical and environmental resource management issues includes: *Digest of Countryside Recreation Statistics* (sport and recreation statistics); *Annual Report of the Countryside Commission* for England and Wales (information on National Parks and Areas of Outstanding Natural Beauty); *Countryside Commission for Scotland Report; Welsh Countryside Committee Annual Report; Nature Conservation Council Annual Report*.

Remote sensing

Since 1960, a new and exciting source of data has been available to geographers. The launch and operation of the first polar orbiting NOAA satellite in 1960 heralded the start of a series of research programmes which has put almost one satellite per year into orbit for mapping purposes. The sensing equipment has improved substantially since the early pioneering satellites, and many now carry a range of scanners capable of sensing a variety of visible and infra-red wavebands as well as radar images which, with improvements in resolution, allow features as small as 20m across to be seen. A summary of the most useful satellites is given in Table 3.10, which also provides information on the number launched and the type and resolution of the image available. An example of the type of image available for Meteosat is given in Figure 3.12 and from Landsat II in Figure 3.23. Further details relating to the availability of national and international remotely sensed data can be obtained from the address given in the margin.

The Remote Sensing Centre,
Space Department,
Royal Aircraft Establishment,
Farnborough,
Hampshire,
GU14 6TD.

**TABLE 3.10
Satellite images**

Satellite	No. Launched	Earliest	Latest	Sensor type	Best Resolution
LANDSAT	5	1972	1984	MSS	30 m
				TM	80 m
SPOT	1		1985	HRV	20 m
				New IR	20 m
NOAA**	9	1960*	1984	AVHRR	1 km
METEOSAT†	2	1977	1981	V	2.4 km
				IR	5.0 km

MSS Multi Spectral Scanner
TM Thematic Mapper
HRV High Resolution Visible
IR Infra red
AVHRR Advanced Very High
Resolution Radiometer
V Visible

* TIROS 1 satellite
** POLAR Orbiting
† Geostationary on equator 0° longitude

**Figure 3.23
A Landsat photograph of the Nile
delta**

It is not the purpose of this section to illustrate how data should be analysed or presented graphically, since these aspects are treated in detail in the remainder of the book. The exercises outlined below are designed to help you explore the availability and application of secondary data to geographical problem solving.

EXERCISE 1:

SOCIAL GEOGRAPHY OF THE NINETEENTH CENTURY CITY

TABLE 3.11 A scheme for the social stratification of household heads from Census Enumerators' books

CLASS	QUALIFICATIONS
I	Heads whose households 1 employed more than 25 people 2 contained at least one servant per household member Heads of professional occupation whose households contained at least one servant per three household members
II	Heads whose households 1 employed between one and 24 people 2 contained at least one servant per three household members Heads of professional households
III	Heads whose households contained servants Heads of non-manual occupation, including those engaged in commerce
IV	Heads of skilled manual occupation
V	Heads of unskilled manual occupation

Skilled occupations involved training, some degree of responsibility or special talent.

The Censuses of 1841 to 1881 provide valuable information on individuals and their households. By plotting the distribution of selected characteristics drawn from the Census enumeration books an insight will be gained of the social geography of the early modern city. Of particular interest is evidence which relates to the social class composition of individuals and their place of residence. A simple way of allocating individuals to a social class is shown in Table 3.11.

Choose a Census for study. A full cover of the urban population is not feasible and so a sample should be taken. Systematic sampling offers a straightforward technique, whereby a regular sequence of households is selected. If the city is very large you will have to select parts of the city to sample. Usually only the head of household is included in the survey. In order to gain an impression of the distribution of social classes it is convenient to combine groups I and II (high social class) and groups IV and V (low social class).

Search the archives for an appropriate street map drawn around the time of the Census. At the beginning of each Census Enumerators' Return there will be a description of the extent of the district in which data were collected. This description will enable you to complete a map to show the enumeration districts. Accordingly, it may be possible to analyse the data at three different scales: by individual household; by street; by enumeration district.

Consider the location of high and low social classes. Do any patterns stand out? Do these patterns coincide with the physical geography of the city? Compare these nineteenth century distributions with those predicted by modern urban land use models. What differences are found?

EXERCISE 2:

URBAN DEPRIVATION

Unlike the nineteenth century Census the modern Census provides details only about areas. An interesting project is the use of the Small Area Statistics of the Census to determine the extent of deprivation and its spatial impact on an urban area.

Ⓠ ☐1

TABLE 3.12
Deprivation indicators

The first step is to chose a set of variables available in the Census which inform about deprivation. It is useful to arrange these into broad categories. A scheme is suggested in Table 3.12.

HOUSING	SPECIAL NEEDS	HOUSING TENURE
Lack bath or inside w.c. >1.5 persons per room (overcrowding) Share bath or inside w.c.	Lone parent households Pensioner households Household heads born in New Commonwealth or Pakistan Children 0–4 divided by women 16–44 (fertility rate)	Council accommodation Rented accommodation
EMPLOYMENT Unemployed S.E.G.11 (unskilled workers)		**ASSETS** No car

[2] In the next stage examine each enumeration district (ED) in turn and record the incidence and extent of these indicators. For example, how many households in an ED lack a bath or inside w.c.; how many households are overcrowded and so on.

[3] The third part of the exercise involves working out a deprivation score for each enumeration district. This can be achieved by scaling each indicator in relation to the data set. For example, if there are 20 EDs in the town, each indicator can be scaled from 1 for the worst ED, to 20 for the best ED. By totalling the scaled indicator scores a cumulative deprivation value is obtained for every ED.

Consider the range of deprivation scores, are there any EDs which stand apart from the others? Plot the distribution of the deprivation scores. Which areas are associated with the highest and lowest scores? Are any patterns evident? Some EDs may score highly on only some of the indicators – which are these?

EXERCISE 3: UNDER THE WEATHER

Weather forecasting is an everyday part of life in Britain and we can often get the information we need from newspaper reports and broadcast national news bulletins. Imagine your misery when you find yourself leading an Outward Bound expedition to a remote part of Britain and you have no access to newspapers or television and the nearest telephone is 20 km away. All you have available to you is a radio capable of receiving the BBC long wave shipping forecast and some copies of the sea area map and booking sheets given in Figure 3.17. Your job is to listen to the shipping forecast and decide whether you are to lead a walking party into the mountains of the Lake District for the next 12 hours. You are responsible for ensuring that your party is clothed appropriately for the weather conditions and all members are correctly equipped.

Q [1] Listen to a shipping forecast (record it if you cannot get the information down fast enough) and transfer the information to the map. Try to mark on areas of high and low pressure and plot where you think the isobars should be from the coastal station information. Use appropriate meteorological symbols to record wind speed and direction for each of the coastal stations. To satisfy the restless members of your party, you should present them with the weather map and your interpretation of what the map means for the following day's expedition. Remember to keep a copy of your map and check your forecast against weather maps published in the following day's newspapers when you return to 'civilisation'. This part of the exercise is particularly useful and can be used to write a self-evaluation of your skills as a forecaster.

EXERCISE 4: A KNOTTY PROBLEM

Examine the data given to you in Table 3.13, relating to deforestation in tropical countries between 1981 and 1985. You are to write a report summarising these statistics with a view to predicting the outcome of deforestation on such a scale and recommending why and how the rate of deforestation should be slowed down significantly in the future. However, your analysis must be placed in context with data from other areas of the world, especially in the UK. You are required to answer the following specific questions:

Q [1] Find out why rates of deforestation are so high in these countries. (Is all the timber burned or used locally, or sold on the international market. If the latter is correct, to whom is it sold?)

61

2 Which countries of the world have the highest and lowest proportions of their total area under natural forest? (Exclude arid countries and high latitude countries from this list.)

2 Which countries of the world have the highest and lowest proportions of their total area under natural forest? (Exclude arid countries and high latitude countries from this list.)

3 Which countries of the world have the highest and lowest proportions of their total area under plantation forest? (Exclude the same countries here.)

4 With reference to the UK, and to your own local area if possible, find out how much natural woodland remains in your county and how much has been lost in the last 100 years or so.

5 If little has changed in the last 100 years, try to establish when most of the county was cleared of forest.
When the county was being cleared, what was the highest rate of deforestation?

Use all of these various data from published statistical maps and historical reviews to analyse the deforestation problem.

TABLE 3.13 Deforestation in tropical countries, 1981–85

Country	Closed forest area, 1980 (thousand hectares)	Annual rate of deforestation 1981–85 (percent)	Area deforested annually (thousand hectares)	Country	Closed forest area, 1980 (thousand hectares)	Annual rate of deforestation 1981–85 (percent)	Area deforested annually (thousand hectares)
Group I[a]				Zaire	105975	0·2	182
Colombia	47351	1·7	820	Cameroon	18105	0·4	80
Mexico	47840	1·2	595	Congo	21508	0·1	22
Ecuador	14679	2·3	340	Gabon	20690	0·1	15
Paraguay	4100	4·6	190				
Nicaragua	4508	2·7	121	**Total**	**979836**	**0·3**	**3160**
Guatemala	4596	2·0	90	**Group III[c]**			
Honduras	3797	2·4	90	El Salvador	155	3·2	5
Costa Rica	1664	3·9	65	Jamaica	195	1·0	2
Panama	4204	0·9	36	Haiti	58	3·4	2
Malaysia	21256	1·2	225	Kenya	2605	0·7	19
Thailand	10375	2·4	252	Guinea-Bissau	664	2·6	17
Lao People's Dem Rep	8520	1·2	100	Mozambique	1189	0·8	10
Philippines	12510	0·7	91	Uganda	879	1·1	10
Nepal	2128	3·9	84	Brunei	325	2·2	7
Vietnam	10810	0·6	65	Rwanda	412	0·7	3
Sri Lanka	2782	2·1	58	Benin	47	2·1	1
Nigeria	7583	4·0	300	**Total**	**6529**	**1·2**	**76**
Ivory Coast	4907	5·9	290	**Group IV[d]**			
Madagascar	12960	1·2	150	Belize	1385	0·6	9
Liberia	2063	2·2	46	Dominican Republic	685	0·6	4
Angola	4471	1·0	44	Cuba	3025	0·1	2
Zambia	3390	1·2	40	Trinidad and Tobago	368	0·3	1
Guinea	2072	1·7	36	Bangladesh	2207	0·4	8
Ghana	2471	0·9	22	Pakistan	3785	0·2	7
Total	**241037**	**1·7**	**4180**	Bhutan	2170	0·1	2
Group II[b]				Tanzania	2658	0·4	10
Brazil	396030	0·4	1480	Ethiopia	5332	0·2	8
Peru	70520	0·4	270	Sierra Leone	798	0·8	6
Venezuela	33075	0·4	125	Central African Republic	3595	0·1	5
Bolivia	44013	0·2	87	Somalia	1650	0·2	4
Indonesia	123235	0·5	600	Sudan	2532	0·2	4
India	72521	0·2	147	Equatorial Guinea	1295	0·2	3
Burma	32101	0·3	105	Togo	304	0·7	2
Kampuchea. Dem	7616	0·3	25	**Total**	**31789**	**0·2**	**75**
Papua New Guinea	34447	0·1	22				

Notes:
a higher than average rate of deforestation and large areas deforested
b Relatively low rates of deforestation but large areas deforested
c High rates of deforestation and small areas of remaining forest
d Low or moderate rates of deforestation and small areas affected

Sources:
1 UN Food and Agriculture Organisation (FAO). 1981.
2 FAO 1981.
3 FAO *Los Recursos Forestales de la America Tropical* (FAO. Rome 1981)
4 FAO/UN Economic Commission for Europe 1985.

Chapter 4 DATA PRESENTATION

4.1 INTRODUCTION

This chapter is concerned with the presentation of geographical data. Data should be displayed in a form which enables the results to be interpreted in a clear manner. Considerable attention should be given to the selection of the most appropriate techniques. The most common ways of presenting geographical data are to use maps, proportional symbols and graphs. Maps and diagrams are important for a number of reasons:

i   there are many occasions when visual representation is the only way of *illustrating a spatial concept;*

ii they help *summarise data* which if written would require several pages of text to describe;

iii good illustrations help *clarify written text;*

iv visual data often *impresses points more forcibly* than if the same data were expressed verbally or in a tabular form;

v   in any field study maps need to be included simply to *locate places* mentioned in the text;

vi well-designed drawings make a project more *attractive* to the reader.

4.2 DRAWING MATERIALS AND EQUIPMENT

The drawing of any illustration requires the use of various drawing instruments and items of equipment. For the student who has only a limited amount of mapping to do many of the routine draughting operations can be carried out with a small basic set of materials.

A recommended set of instruments and materials

Technical drawing pens, nib sizes 0.35mm and 0.50mm

International standard stencils, 0.35mm and 0.50mm

HB and 2H pencils

Indian ink

Flexible curve

45° set square

Ink and pencil springbow and larger compasses

Ruler and protractor

Set of graph paper of different grid sizes, including circular graph paper

Tracing paper

Masking tape for fixing sheets of paper to a drawing board or table and which can be removed without damaging the surface of the drawing

Razor blades for erasing inked lines; soft rubber for removing pencil

Map pens: Technical drawing pens are the most useful all-round pens for mapwork. They look like fountain pens but have a tubular point rather than an ordinary nib. These pens are produced under several trade names such as Rotring (Rapidograph, Isograph or Variant), Mars Technical Pen and Standardgraph. A nib size of 0.50mm will suffice for most purposes, although it is useful to have others for finer and thicker lines, such as 0.35mm to 0.70mm.

	16pt	12pt	8pt

Futura medium

A B C D
a b c d
1 2 3 4

A B C D
a b c d
1 2 3 4

A B C D
a b c d
1 2 3 4

Univers 65

A B C D
a b c d
1 2 3 4

A B C D
a b c d
1 2 3 4

A B C D
a b c d
1 2 3 4

Helvetica medium

A B C D
a b c d
1 2 3 4

A B C D
a b c d
1 2 3 4

A B C D
a b c d
1 2 3 4

Figure 4.1
Dry transfer lettering styles and sizes

4.3
MAP DESIGN AND LAYOUT

Ink: For normal drawing a jet-black Indian ink such as Pelikan or Rotring is recommended. This provides a clear, durable, consistent image which will enhance any project.

Lettering: Hand-lettering must be done well or not at all. Useful aids are stencils or transfer lettering sheets, such as Letraset. Stencils are very convenient and relatively cheap. Practice is needed to get the correct alignment and spacing of letters. Stencils should always be used against a straight edge, that is, a set square or ruler, with the pen held vertically. Ensure that the nib size is compatible to the stencil template.

Dry transfer lettering sheets, although more expensive, will give a professional finish when set up correctly. Univers, Helvetica and Futura style faces are commonly used. These can be bought in different sizes, but 8, 12 and 16 point are convenient for most purposes (Figure 4.1). Also available are sheets of different shading patterns, textures and numerals.

Paper: Tracing paper is the recommended medium for drawing. Mistakes can be easily erased or scratched out with a sharp blade and the paper smoothed again with the back of a finger nail. If an attempt is made to draw in ink over an erased area which has not been burnished in this way a clean, sharp line will not be produced.

Graph paper is essential for most cartographic work as it can be used under tracing paper to:

i construct map frames at right angles;
ii arrange dots and symbols regularly by placing them at grid intersections;
iii line-up hand shadings;
iv place lettering parallel to the top and bottom edges of the map or diagram.

A well-designed illustration should be clear, balanced, not over-loaded with information, and self-explanatory.

Marginal information
All the information required to understand a map should be found around its margins. Marginal information includes a *title, scale, north-point, key* and *source of data.*

There are three common expressions of scale: *verbal, representative fraction (RF)* and *linear* (Figure 4.2).

Verbal	Representative fraction
One inch equals 1 mile	1:63 360
2 centimetres to 1 kilometre	1:50 000
About 4 miles to 1 inch	1:250 000

Figure 4.2
Methods of representing scale on maps

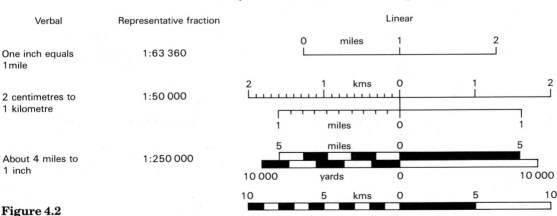

A linear scale is suitable for most project maps. Like the north-point, its form should be simple and not elaborately drawn.

Any symbol used on the map must be placed in the key. It is unnecessary to print *Key* above a legend or to include *A map of* . . . . . in the title.

Layout
A successful map starts with layout. It is useful to provide a border around a map. The map should be proportioned so that it is as large as possible within this frame (Figure 4.3). A good guideline is to *minimise white space on artwork.*

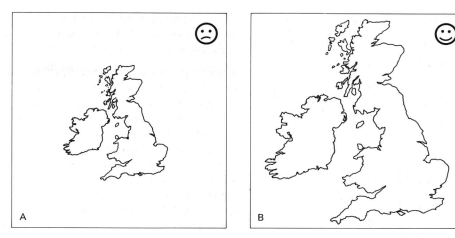

Figure 4.3
The map should be proportioned so that it is as large as possible within its frame, as in map B

Marginal information needs to be economically arranged *within* this frame. The following points should be considered when working out a satisfactory layout (Figure 4.4):

i make a number of layout sketches, do not be satisfied with the first attempt;

ii avoid top-heavy and side-heavy arrangements, try for a balanced composition (A and B);

iii treat the study area as a feature that can be skewed so as to make the best use of space (C and D);

iv do not reduce the study area to create free spaces at its edges. Instead keep the map as large as possible and look for alternative areas to position marginal information (E and F).

Figure 4.4
Preferred ways of arranging marginal information

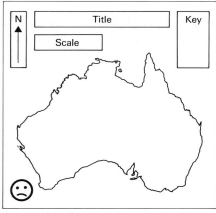

A   Avoid top heavy arrangements

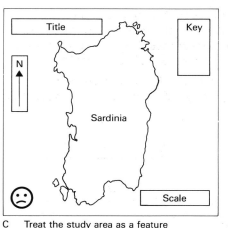

C   Treat the study area as a feature that can be skewed

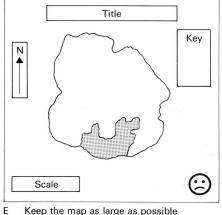

E   Keep the map as large as possible

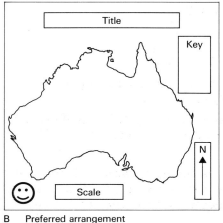

B   Preferred arrangement

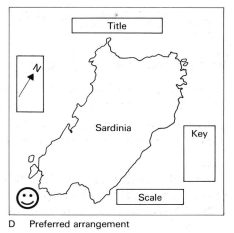

D   Preferred arrangement

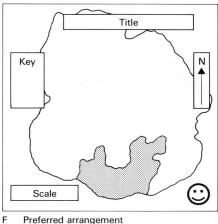

F   Preferred arrangement

Verbal information

Some general guidelines (Figure 4.5):

i in order to avoid visual clutter when placing names on a map, slightly exaggerate the spacing between letters (A and B);

ii names should not be bisected by linework, the part of the line crossed by the name should be removed (C and D). Wherever possible, place the name to one side of the symbol to which it refers and slightly above or below its centre (E and F);

iii places on the east bank of a river have their names to the east of the symbol, those on the west bank have their names to the west (G and H);

iv rather than spread the name of a linear feature, such as a road or river along its length, repeat the name at intervals (I and J);

v places on the coast usually have their names placed at an appropriate angle in the sea (K and L);

vi fully-enclosed features have their names contained wholly within the feature (M and N).

Spacing of letters: In order to achieve a visually pleasing map it is important to space and arrange correctly the individual letters which make up a word. Some useful conventions are shown in Figure 4.6.

Figure 4.5
Verbal information: some general guidelines

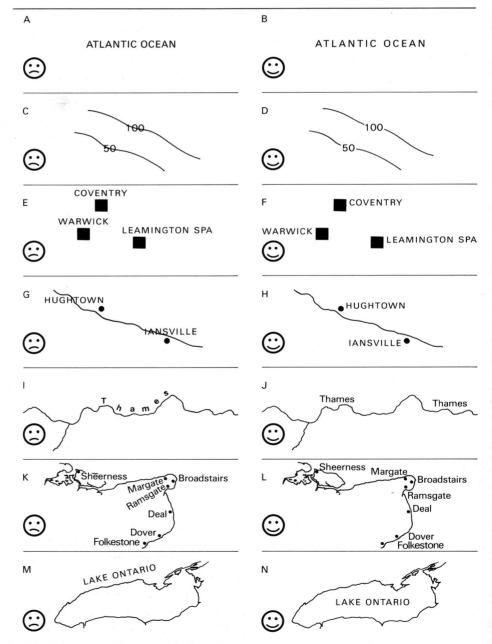

Figure 4.6
Spacing of letters

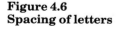

Symbols: points, areas and lines

Features occur at a point, within specified areas, or along a given line. The choice of symbol to represent such data needs careful consideration. When designing symbols there are some useful points to bear in mind:

i the ideal symbol should be recognisable without reference to a key;

ii every symbol should be distinctive and distinguishable at a glance;

iii symbols should be appropriate. Rather than assigning sets of symbols arbitrarily to different data, consider what features are to be shown. For example, it is conventional to represent a church by a cross, boundaries as solid or broken lines, or areas of limestone in a familiar 'brick-wall' manner;

iv symbols should be kept as small as possible to avoid dominating the map. However, it will rarely be possible to draw symbols according to the scale of the map. An obvious example is a road which on any small-scale map occupies much more than its true width;

v when designing symbols it is sensible to ensure that they are easy to draw and repeat. The same symbol needs to be of the same size and shape as the others;

vi symbols should be drawn with clean lines, good open spaces and square corners where necessary. Stencils and dry-transfer sheets are valuable drawing aids. Graph paper used underneath tracing paper provides a useful framework for different shading textures and patterns.

4.4 MAPPING GEOGRAPHICAL PATTERNS

A classification of map types is shown in Figure 4.7. The three rows of the scheme show levels of measurement and the three columns reflect how data can be gathered and presented. The result is nine cells which contain a range of map types. The classification also recognises three broad groups of maps: those showing qualitative and quantitative data and an intermediate set showing ordinal (ranked) data.

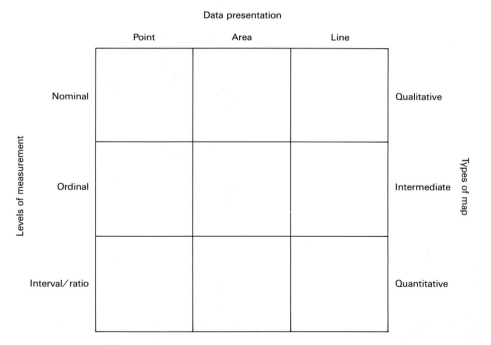

**Figure 4.7
A classification of map types**

4.5 QUALITATIVE MAPS

These maps *locate* data measured on a *nominal scale*. The information that they show *distinguishes different categories of phenomena*, with no indication of quantities, amounts or relative value (see Chapter 1).

Points

The simplest kind of symbol is a dot. A place either has or has not a particular feature and a dot represents its presence. Other types of geometrical symbol can be used such as squares, triangles, half-circles and hexagons. Alternatively, simple representations of real objects may be

Figure 4.8
Point symbols: examples from the list in International Yearbook of Cartography, Vol 6

INDUSTRY

Electricity production		**Mechanical industry**	
Atomic energy		Foundry products	
Hydraulic energy		Nuts, bolts, screws	
Gas production		Agricultural machinery	
Petroleum refinery		Railway equipment	
Coal		Motor cars	
Iron		Shipbuilding	
Non-ferrous metals		**Chemical industries**	
Petroleum		Mineral chemical industry	
Other extractive industry		Rubber	
Metallurgy		Photography	
Iron and steel production			

THE COUNTRYSIDE

Panorama		Viewpoint	
Waterfall		Caves	
Conifers		Deciduous	
Palm trees		Spring	
Medieval ruins		Other ruins	
Fortifications		Castle	
Fortress		Chateau, manor house	
Church, cathedral		Abbey, monastery	
Mosque		Synagogue	
Dam, barrage		Bridge or viaduct	
Lighthouse		Telecommunications	
Vineyard		Wine cellar	
Postal service		Telephones	
Customs post		Bus station	
Car ferry		Railway station	
Boat service		Telepherique	
Civil aerodrome		Hospital	
Hotel and number of beds		Golf course	
Mountain hut		Beach	
Fishing port		Salt marsh	

preferred, like an aeroplane to show the location of an airport or a boat to show the presence of a ferry (Figure 4.8).

Areas

These maps distinguish one kind of area from another, without attempting to show its value. They include maps of land use, farming types, soils, vegetation and geology.

Simple shading patterns or textures, colours and pictorial representations can be used to symbolise such data (Figure 4.9). Since quantities are not being shown, categories should not be distinguished by increasing darkness, as this strongly suggests differences in quantity. Instead, it is better to try to hold darkness constant.

On maps showing *one* kind of information, for example, types of hardwood plantation, categories should be distinguished by *one* form of symbol, like shading patterns. When two or more kinds of information are being shown on a map, for example, types of hardwood and softwood plantation or regions of erosion and deposition, it is appropriate to distinguish these groups by using different forms of symbol, like shading patterns and textures.

Mapping physical regions in Europe provides an example. Figure 4.11(a) shows a presentation that is acceptable. Each type of physical region has been given a distinctive pattern which can be identified in the key. Closer scrutiny of the data reveals two broad groups of physical region, mountain systems, and plateaus and plains. In Figure 4.11(b) these broad groups have been distinguished by different kinds of patterns *and* textures. Differences within each group are still clear, but, in addition, the map areas occupied by each broad group stand out.

Lines

Nominal line data include the representation of rivers, coastlines, boundaries, roads and railways. These lines show the nature of the feature without any indication of status. They may be used for many purposes. For example, a line portraying a road may show its exact route on a map or, at a smaller scale, may reveal that two places are connected by a road. If more than one type of linear feature is represented on a map each must be clearly distinguished, usually by means of size, shape and colour (Figure 4.10).

Figure 4.9
Methods of distinguishing one kind of area from another

Figure 4.10
Line symbols: by varying the nature and weight of line a wide variety of symbols can be achieved

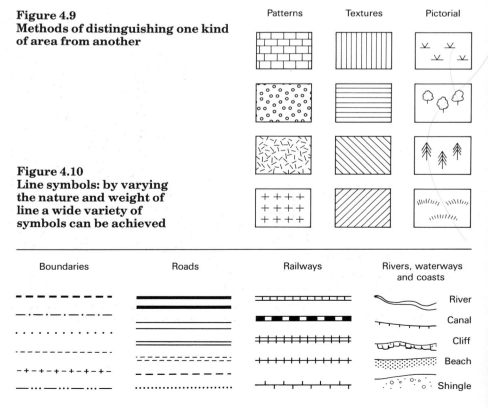

**Figure 4.11
Mapping physical regions of
Europe using different kinds of
patterns and textures**

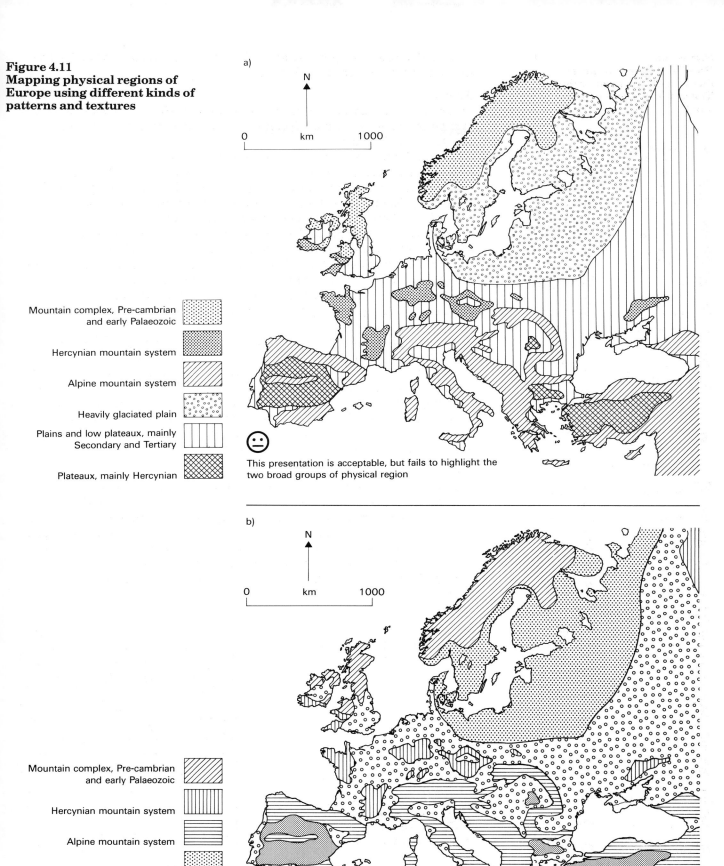

a)

N

0   km   1000

Mountain complex, Pre-cambrian
and early Palaeozoic

Hercynian mountain system

Alpine mountain system

Heavily glaciated plain

Plains and low plateaux, mainly
Secondary and Tertiary

Plateaux, mainly Hercynian

This presentation is acceptable, but fails to highlight the
two broad groups of physical region

b)

N

0   km   1000

Mountain complex, Pre-cambrian
and early Palaeozoic

Hercynian mountain system

Alpine mountain system

Heavily glaciated plain

Plains and low plateaux, mainly
Secondary and Tertiary

Plateaux, mainly Hercynian

This presentation is preferred; two broad groups of physical
region stand out and differences within

This is an exercise in qualitative mapping. It combines the use of point, line and area symbols to distinguish different features of a landscape.

EXERCISE 1: POINTS, LINES, AREAS

[Q] [1] Figure 4.12 shows an imaginary island called Davian. Trace the outline of this island. In black ink, complete the map of the island by adding to the tracing details about point, line and area phenomena given in the written description below. Remember to provide full marginal information.

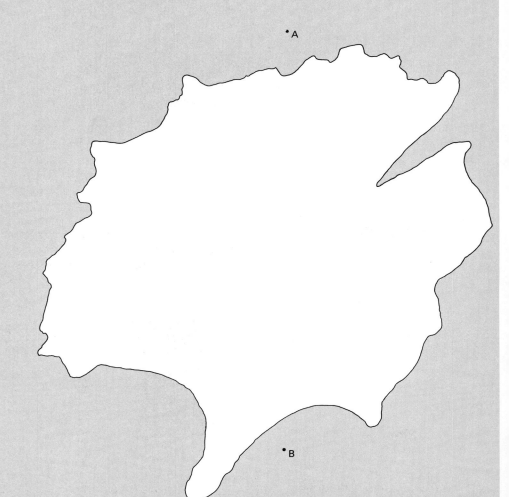

Figure 4.12
Outline of the island of Davian

Davian has been recently surveyed by the Fomat Institute. From point A to B is 8 kilometres.

The northern coastline is bordered by steep cliffs which are flanked by a narrow shingle beach. In contrast, in the extreme south-west and south-east of the island two wide sandy bays converge onto a small peninsula. Both bays are backed by an area of low-lying sand dunes and marsh. Elsewhere the coastline is undistinguished, consisting of sand and shingle interrupted by wave-cut platforms.

In the centre of the island, rising in a highland region of 600 metres, the River Cherry flows north-eastwards out into the sea. A number of tributaries flow into the river along its course.

Near the mouth of the river, straddling both its banks, is Libbyville, the major town and port of the island. Libbyville is a ferry terminal and has a small airport on its southern margins. The town's water supply is provided by a reservoir located about 2.5 kilometres in a south-southwest direction from its centre. On the southern peninsula of the island is the fishing village of Lillington, which has a lighthouse on its small, man-made harbour. The only other area of settlement is found in the low-lying area bordering the west coast; this is a region of scattered farmsteads.

The main economic activities of the island are cider-making, forestry and fishing. Much of the farming area is given over to apple orchards. In the north of the island extensive coniferous plantations have been established.

Lillington is linked by road and single-track railway to Libbyville. The road follows a route along the east coast, whereas the railway takes an inland course with embankments and cuttings. Both enter the town from the south-east and cross the mouth of the river by separate bridges. Most of the farmsteads in the west of the island are located near or on minor roads leading to Libbyville. In the northern region forest tracks join onto a road leading to the main town. The old mineral railway line which served this area is now disused. A coastal footpath girdles the island.

The island is divided into three parishes, Libbyville, Lillington and the northern and western parish of Outerbury. The parish boundaries meet at a point near the headwaters of the River Cherry.

4.6
INTERMEDIATE MAPS

Maps which portray *ordinal data* fall into an intermediate position between qualitative and quantitative maps. These maps show *location* and the *relative status, importance* or *ranked position* of phenomena.

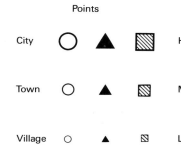
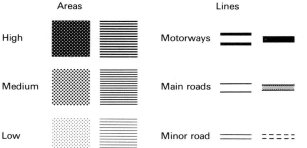

**Figure 4.13
Methods of representing ordinal data**

Points
By carefully varying the colour, size and shape of symbols the relative importance of point features can be shown on a map. Some different ways of distinguishing the relative order of settlements are shown in Figure 4.13.

Areas
In this category, colours, patterns and textures can be used to distinguish one area from another. The higher ranks or greater amounts are usually depicted by darker shades or greater brilliance of colour (Figure 4.13). There is no need to grade classes; simply ensure that the shading or intensity of colour changes from light to dark. Common examples of these sorts of map include regions of advanced, intermediate and low industrial development or areas of high, medium and low rainfall.

Lines
The relative importance of line features can be shown by differing line widths or by varying colours and symbols. Figure 4.13 shows some of the ways in which roads can be distinguished.

These three exercises are all concerned with mapping ordinal data. In each case you are asked to organise and arrange the data into categories and to represent these groups by point, line or area symbols. The dispersion diamgram (Figure 4.30) is useful for grouping data.

EXERCISE 1: POINT DATA

Q 1   Classify the data shown in Figure 4.14 into three groups to represent high, medium and low levels of zinc concentration in mussels.

2   Devise suitable point symbols to portray this information and plot these symbols on a tracing of Figure 4.15.

3   Attempt to explain the principal variations in the levels of marine pollution around the English and Welsh coastline.

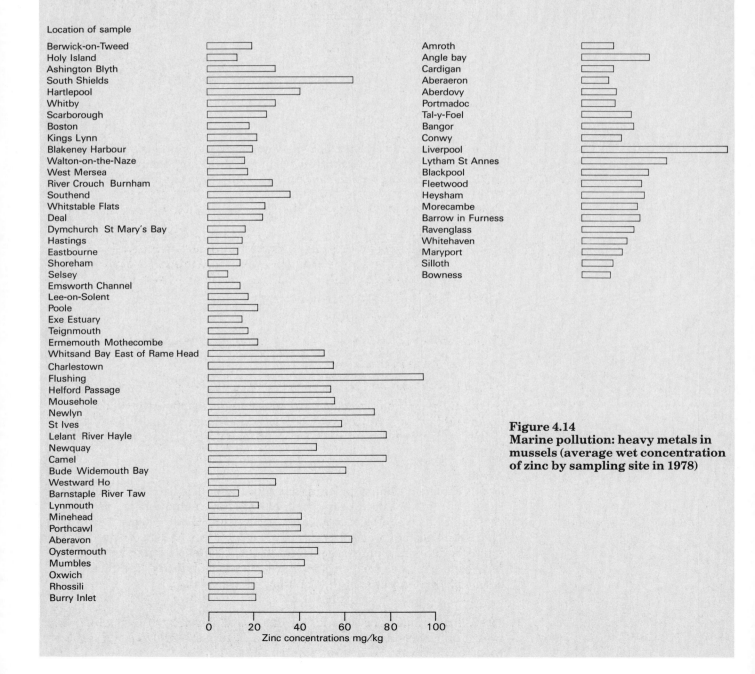

Figure 4.14
Marine pollution: heavy metals in mussels (average wet concentration of zinc by sampling site in 1978)

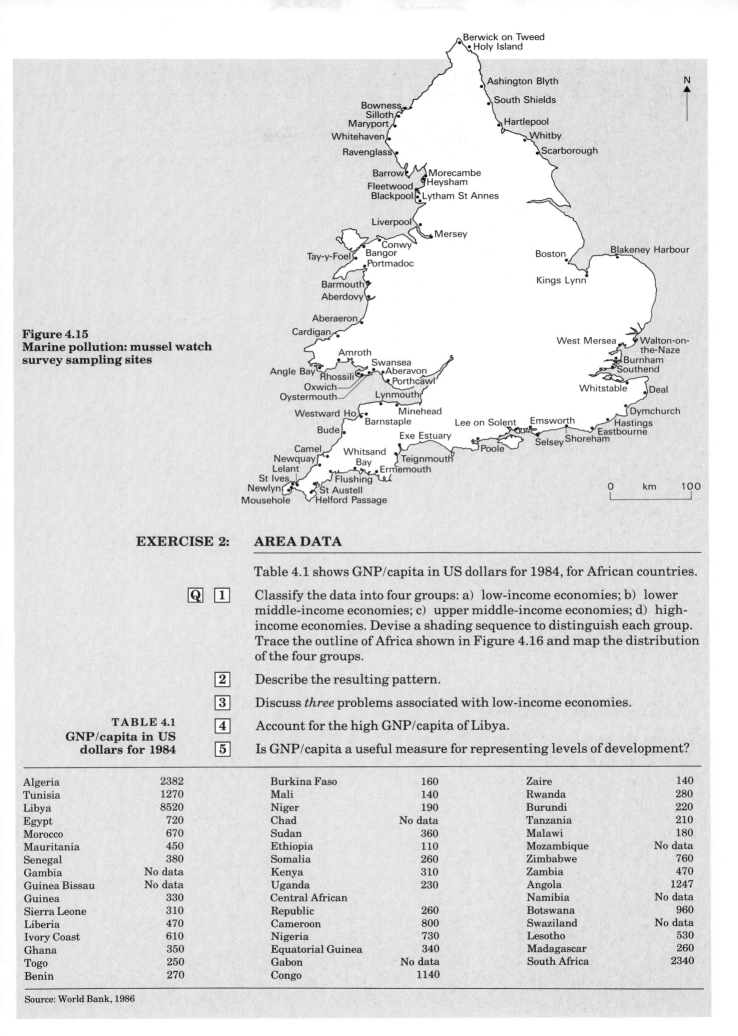

Figure 4.15
Marine pollution: mussel watch survey sampling sites

EXERCISE 2: **AREA DATA**

Table 4.1 shows GNP/capita in US dollars for 1984, for African countries.

Q 1 Classify the data into four groups: a) low-income economies; b) lower middle-income economies; c) upper middle-income economies; d) high-income economies. Devise a shading sequence to distinguish each group. Trace the outline of Africa shown in Figure 4.16 and map the distribution of the four groups.

2 Describe the resulting pattern.

3 Discuss *three* problems associated with low-income economies.

4 Account for the high GNP/capita of Libya.

5 Is GNP/capita a useful measure for representing levels of development?

TABLE 4.1
GNP/capita in US dollars for 1984

Algeria	2382	Burkina Faso	160	Zaire	140
Tunisia	1270	Mali	140	Rwanda	280
Libya	8520	Niger	190	Burundi	220
Egypt	720	Chad	No data	Tanzania	210
Morocco	670	Sudan	360	Malawi	180
Mauritania	450	Ethiopia	110	Mozambique	No data
Senegal	380	Somalia	260	Zimbabwe	760
Gambia	No data	Kenya	310	Zambia	470
Guinea Bissau	No data	Uganda	230	Angola	1247
Guinea	330	Central African		Namibia	No data
Sierra Leone	310	Republic	260	Botswana	960
Liberia	470	Cameroon	800	Swaziland	No data
Ivory Coast	610	Nigeria	730	Lesotho	530
Ghana	350	Equatorial Guinea	340	Madagascar	260
Togo	250	Gabon	No data	South Africa	2340
Benin	270	Congo	1140		

Source: World Bank, 1986

73

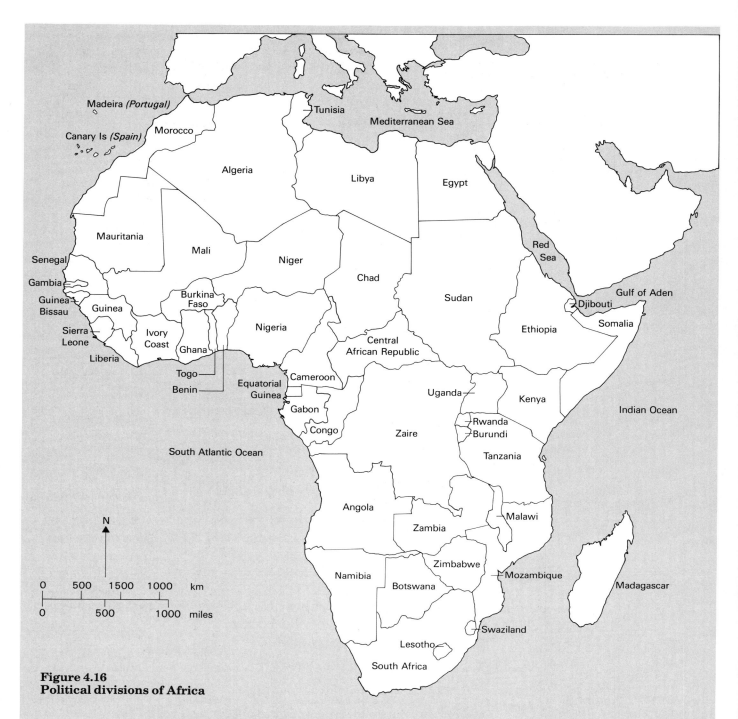

Figure 4.16
Political divisions of Africa

EXERCISE 3: LINE DATA

Table 4.2 shows 18 of the 529 gauged British rivers.

Q 1 Rank each river from high to low according to a) its gauged length;
b) gauged catchment area; c) mean annual discharge.

2 Combine the ranks for each river and classify the rivers according to the
sum of their ranks.

3 Devise an appropriate linear symbol to represent each category. Plot
these symbols on a tracing of Figure 4.17.

4 Justify your groups and describe the characteristics of each group of
rivers.

5 What factors are likely to influence the mean annual discharge rates of
rivers?

TABLE 4.2
Some characteristics of selected major British rivers

River	Length (km)	Catchment Area (km²)	Mean annual discharge in cubic metres/second	River	Length (km)	Catchment Area (km²)	Mean annual discharge in cubic metres/second
Aire	114	1930	36.89	Spey	137	2650	55.86
Avon	125	2210	14.43	Tay	110	4590	152.21
Clyde	105	1700	37.40	Tees	103	1269	19.46
Dee	116	1370	35.70	Thames	239	9950	67.40
Eden	102	1370	31.02	Trent	149	7490	82.21
Great Ouse	184	3030	14.16	Tweed	140	4390	73.85
Ouse	117	3320	40.45	Tyne	87	2180	43.45
Ribble	94	1140	31.72	Tywi	82	1090	38.34
Severn	206	4330	62.70	Wye	225	4040	71.41

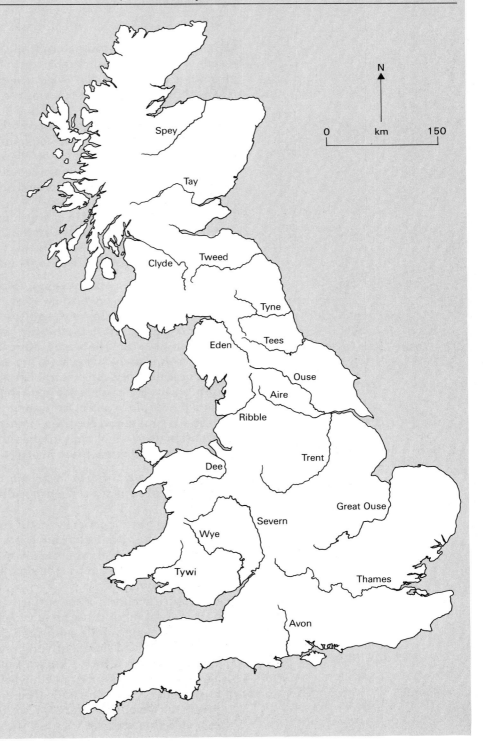

Figure 4.17
Major British rivers

The main purpose of quantitative mapping is to show *quantity* as well as *kind* and *location*. These are maps which represent *interval* and *ratio* data. This category is by far the most important of the three and includes a range of popular thematic maps. Like non-quantitative maps, data are represented as points, areas and lines.

Points

There are two major types of quantitative point data:

i *continuous data:* phenomena which occur everywhere but are collected only at convenient points, such as the height of land and atmospheric pressure;

ii *non-continuous (discrete) data:* phenomena which are tied to specific locations or relate to particular places, such as mineral production and populations of towns.

Each of these categories will be discussed in turn.

Continuous data: isoline mapping. Maps which show amounts by means of lines plotted through points of equal value are known variously as *isolines, isopleths* or *isarithms*. The prefix *iso* is derived from Greek and means the *same*. The most familiar use of *isolines* is to connect points of equal height above sea-level, in this case they are known as contours. Other isoline maps include *isotherms* for temperature, *isobars* for atmospheric pressure, *isohyets* for rainfall and *isochrones* for time-distance data. These lines are used to show change from place to place and gradients of change. Where lines are crowded together, rates of change are sharp, where they are more widely spaced, variation is gradual.

Three stages are involved in isoline mapping:
  i placing the values from which the lines will be drawn;
 ii setting the values of the isolines;
iii drawing in the isolines by *interpolation*.

Placing the values. Isoline mapping depends on a sample of point data taken from what can be considered as an infinite number of possible points. Ideally, as many stations or control points as possible should be used as a basis for collecting the data, the aim being to provide a comprehensive and even coverage of the whole area. More usual is that point data are irregularly and unevenly spaced. For example, surveyors mapping the form of land are only able to establish spot heights at intervals across a terrain, weather data depends upon the location of weather stations, and time-distances often reflect the spacing of settlements. As a result, isoline maps rarely provide an accurate picture of a distribution, as the information upon which they are based is incomplete; at best they show a *three dimensional representation* of a statistical surface.

Setting isoline intervals. On the whole a uniform interval gives the most easily interpreted picture. When deciding upon the value of the interval consider:

i the scale of the map. Compare the contour intervals on OS 1:25 000 and 1:50 000 maps. On the former, contours are drawn at every 25 feet (to the nearest 1 metre), whereas on the larger scale map a 50 foot interval is used. A rule of thumb is the smaller the scale of a map the finer the isoline interval;

ii the nature and form of the surface. Intervals should be chosen to bring out points of significance about a distribution. Look at the overall spread and value of the points and consider whether the interval which has been selected sufficiently highlights patterns of variation. If the shape of a land surface is known, ensure that the contour interval is able to represent important changes in topography;

iii the number of sample (control) points. Generally, the fewer control points the fewer the number of isoline intervals. Drawing many isopleths from data for very few points will obviously give a false impression of accuracy.

Drawing isolines. Isoline mapping is more than just a matter of joining the points of the same value. Rarely will isopleth intervals coincide with the value of control points. Instead the lines are drawn by assessing by eye the position they pass between two control points. This process is known as *interpolation*.

A basic assumption of interpolation is that the gradient between points of known value is constant. Figure 4.18 shows how to locate an isoline between two control points. In practice, it is not necessary to carry out this full procedure between every pair of points, as careful estimation by eye will offer a satisfactory result.

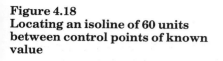

Figure 4.18
Locating an isoline of 60 units between control points of known value

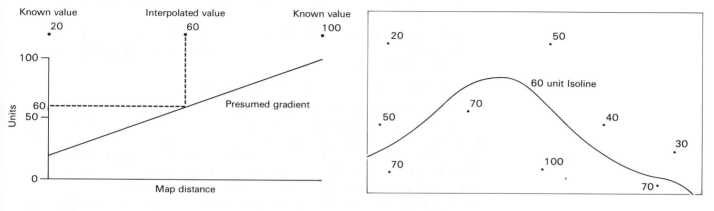

Other guidelines include (Figure 4.19):

i an isoline or contour never splits (A and B);

ii all isolines which occur between two points should be drawn, even if the rate of change is very steep. Lines should not be missed-out simply for convenience. On topographic maps it is conventional to show sharp breaks of slope, such as cliffs, by means of symbolisation and these symbols can be used to interrupt isolines (C and D);

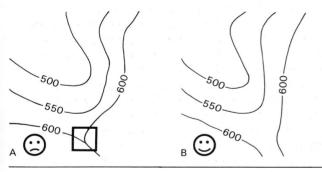

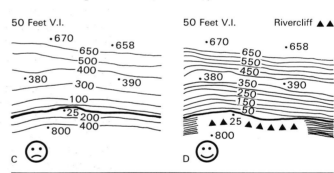

Figure 4.19
Isoline mapping: useful guidelines

iii isolines should be clearly labelled and the vertical interval shown in the key. Isoline labels should break the lines and should be aligned with them and not with the map frame (E and F);

iv not all isolines need to be numbered, but visually it must be clear whether lines are increasing or decreasing in value and the magnitude of these changes must be apparent. Isoline mapping is often easier if a

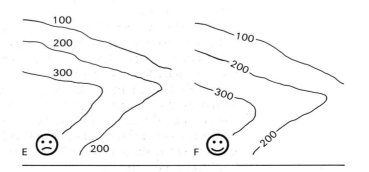

system is followed. Start with the highest or lowest areas and define their extent. Then trace how the lows give way to highs and in this way the general direction of values will be correctly placed;

v   geographical interpretation of data is essential. For example, on a contour map ensure that rivers flow in valleys and downhill (G and H). Extreme lows are never represented by a single value, instead a pair of lowest value isolines defines the low area or trough (I and J).

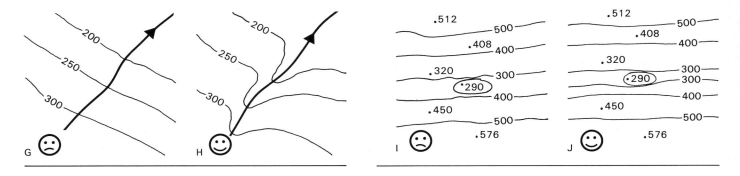

Sometimes shading between isolines is used to help differentiate between lines of different value and to highlight the shape of surfaces. Layer shading is explained in the section on choropleth mapping.

Non-continuous (discrete) data: proportional symbols. These are used to represent the value of phenomena which occur at a number of points. The size of symbol is varied to show change in absolute amounts. Values can be assessed by reference to a scale. Proportional symbols should be compact, easy to draw and capable of estimation by eye. In this type of map the symbol is the most important feature and all other locational detail can be relegated to the background. Care needs to be taken to ensure that the symbol is sufficiently solid to stand out against the background on which it is to be shown. By varying the tone or shading of a symbol mulitiple data can be shown on the same map. For example, energy output can be larger or smaller and produced by electricity, gas or oil. The size of symbols can be proportional to linear, areal or volumetric scales.

Proportional linear symbols: bar graphs. The simplest type of proportional symbol is the bar graph. The length of bar is simply made proportional to the amount being shown. On such a linear scale a bar of 100 units would be twice the length of a bar of 50 units.

Setting a scale. The width of bar and the choice of scale needs to be appropriate to the size of the map and the range of the data. Experiment with different scales until a satisfactory set of bar lengths is achieved. A useful method for devising a scale is shown in Figure 4.20. The horizontal line should be scaled according to the values to be mapped. Draw a bar of convenient height to represent the highest value. Extend a line from the top of the bar to the zero-point. The heights of bars for any values can immediately be seen. If the bar representing the smallest value is too short or if the sizes of other bars cannot be fitted conveniently onto the map, a different length of bar can be drawn for the highest value. The linear scale which is selected must be shown in the key of the map.

Locating the bars. The bases of bars should be drawn near or at the point of data collection, or conveniently located within the region to which it relates. Bars are normally drawn vertically.

Relative merits. Simplicity of construction and ease of interpretation are the chief merits of this type of map. The main disadvantages of bar graphs are the large amount of vertical space required, especially if the range of values is great, and the problem of overlapping symbols occurs.

Variations of bar graphs. If more than one type of point phenomena for a location need to be shown at the same time, bars of different shading can be drawn side by side (Figure 4.21).

Figure 4.20
Bar graphs: devising a scale

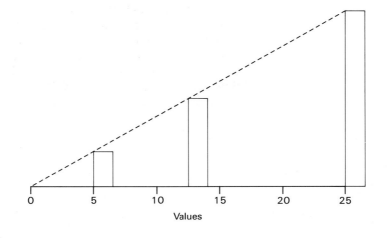

Figure 4.21
Bar graphs showing multiple data

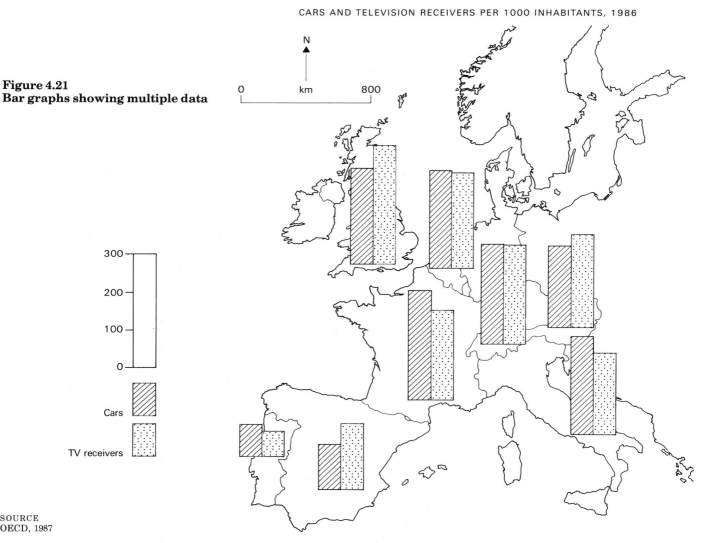

CARS AND TELEVISION RECEIVERS PER 1000 INHABITANTS, 1986

N

0   km   800

Cars

TV receivers

SOURCE
OECD, 1987

Divided bar graphs are useful when attempting to show both total quantity and the value of its components. For example a graph showing total agricultural area can be divided into parts to represent land under dairying, mixed farming and cereal crops. Figure 4.22 shows the average monthly deposition of sulphur in selected European countries and distinguishes the proportion produced internally from that deriving from elsewhere.

Proportional areal symbols: proportional circles and squares. Circles and squares are virtually the only symbols whose areas are made proportional

79

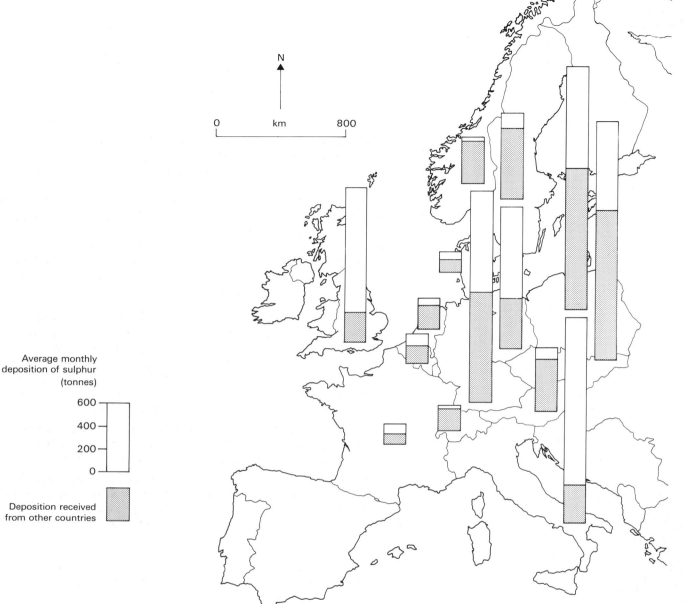

Figure 4.22
Divided bar graphs

DEPOSITION OF SULPHUR IN EUROPE, 1980

N

0   km   800

Average monthly
deposition of sulphur
(tonnes)

600
400
200
0

Deposition received
from other countries

to the values to be mapped. Circles are easier to draw than squares and are
the most popular. Like bar graphs they are used to show absolute values.

Scaling circles or squares. If a map is to be drawn using proportional
circles, the *area* of each circle needs to be proportional to the value shown.
Consider two data values, one twice the other. If the radius of one circle is
made 1 unit and the other 2 units, the area of the circle will not be twice
the first, but four times as great, because the area of a circle equals πr^2
($\frac{22}{7} \times$ radius^2). Therefore, it is necessary to scale the radius of each circle
according to the *square root* of the values to be mapped and not the values
themselves. In the case of squares the lengths of the sides should be
proportional to the square root of the values that they show.

Two methods can be used to devise a suitable scale. The first is a
statistical method:

i   List the square roots of all the values to be mapped.
ii   Multiply each square root by a *constant*. Experiment with various
     constants until a suitable radius for both the largest and smallest
     value is found (Table 4.3).

TABLE 4.3
A method of scaling
proportional circles

Values	Square roots of values	Constant	Radius of circle
980	31·3	0·08	2·5
636	25·2	0·08	2·0
432	20·7	0·08	1·7
100	10·0	0·08	0·8
28	5·3	0·08	0·4

80

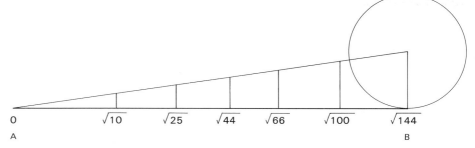

I) Scaling proportional circles: graphic method

$\sqrt{10}$ $\sqrt{25}$ $\sqrt{44}$ $\sqrt{66}$ $\sqrt{100}$ $\sqrt{144}$

Figure 4.23
Proportional circles

II) Methods of overlapping circles

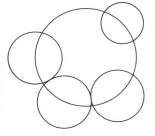

Open circles

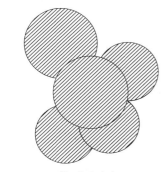

Shaded circles

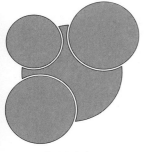

Solid circles

Alternatively, a graphical method can be used (Figure 4.23):

  i List the square roots of all the values to be mapped.
 ii Arrange the square roots along a horizontal line of convenient length (A/B).
iii Draw a circle at the highest square root value (B). The size of this circle should be appropriate to the space this value will occupy on the map.
 iv Extend a line from the centre of this circle to the zero-point (A).
  v Drop perpendiculars from any point along this line to the linear scale (A/B); these represent the radii of circles in proportion to each other.
 vi Experiment with different sizes of circle at the highest value until a convenient set of radii is produced.

Locating the symbols on the map. The proportional symbols should be located as accurately as possible. There are occasions when many circles will need to be drawn in a small space. If the circles are open they can be completely overlapped. If they are shaded they should be stopped where one meets another. Where two black circles overlap a white space can be left where overlapping occurs (Figure 4.23). This overlap should not be considered a flaw: it is a feature of the best-designed maps.

Legend. The value of the proportional symbols must be shown in a key. Two different arrangements are conventionally used: nested circles (Figure 4.24) or a linear scale, such as that shown in Figure 4.23.

Relative merits. The chief advantage of using two-dimensional symbols, such as circles and squares, is that they can represent a large range of data more efficiently than symbols of a linear type. Their disadvantage is that map readers tend to underestimate the size of the larger map circles.

Figure 4.24
Mapping proportional circles

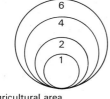

Agricultural area
(hundreds of hectares)

DISTRIBUTION OF AGRICULTURAL AREA IN THE PARISHES OF A COUNTY

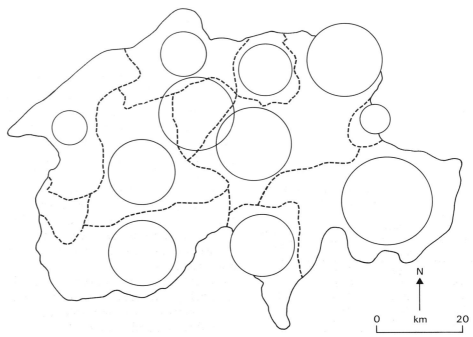

N

0   km   20

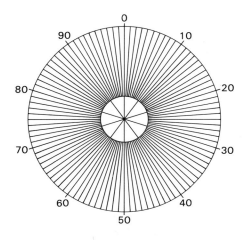

Figure 4.25
Circle divided into hundred parts.
Each part is equivalent to 1 per cent

Representing multiple data: Pie graphs. Proportional circles can be sub-divided into segments to indicate percentages of the total value represented by different components. For example, proportional circles showing the total amount of woodland in British counties can be divided into segments to represent percentages of coniferous and deciduous cover.

Constructing pie graphs. Construction is not difficult since 1% is equivalent to 3.6°. Multiplication of any percentage by this value of degrees produces an appropriate size segment: 10% is equivalent to a segment of 36° and 25% to a segment of 90°.

Plotting the segments. In order to plot these percentages each circle can be divided into a 100 equal parts, with an origin on the radius at a vertical point above its centre (Figure 4.25). There is no need to divide any circle in this manner. A sheet of circular percentage graph paper can be placed under the tracing so that its centre aligns with the centre of a proportional circle. Starting from the origin the required number of segments are measured. Each successive segment is drawn from where another ends. An alternative method is to use a protractor to measure each segment. Once the percentages have been plotted, distinctive shadings can be given to the segments so that components can be compared visually over the map. It is advisable to limit the number of segments to no more than six, otherwise interpretation becomes very difficult. Remember to key the pie graph and its segments (Figure 4.26).

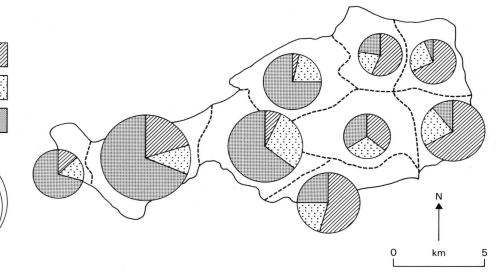

DISTRIBUTION OF SELECTED ETHNIC GROUPS IN THE WARDS OF A CITY

Indians

Pakistani/Bangladeshi

West Indians

Number of people (000s)

Figure 4.26
Mapping pie graphs

Proportional volumetric symbols: proportional spheres and cubes. Where the range of point data are very great, linear and areal scaled symbols may not be able to symbolise both the upper and lower ends of the data set. For example, if the lowest value is 5 and the upper value 50 000 the square roots will be 2.2 and 223.6 respectively. In these circumstances volumetric scaling can better deal with an extreme range.

Cubes and spheres are commonly used to represent volumes (Figure 4.27). The size of the symbols are scaled proportional to the *cube roots* of the data, that is, the sides of a cube or the radius of a sphere. Cubes and spheres can be scaled using the same methods described for areal symbols.

Relative merits. Although volumetric symbols can represent a large range of values, their disadvantage is that map-users tend to underestimate the significance of volume differences. Small variations in the size of spheres often represent large differences in absolute data. Also, for the student, these symbols are difficult to draw.

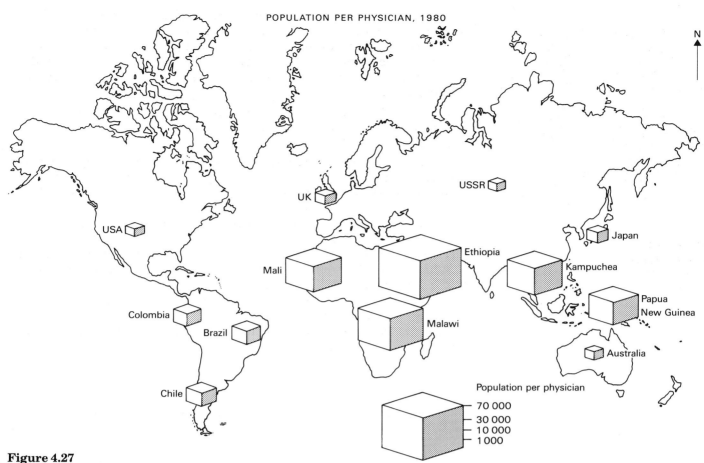

Population per physician

70 000
30 000
10 000
1 000

Figure 4.27
Proportional cubes: the cubes are
proportional to
$$\sqrt[3]{\text{population/physician}}$$
(multiplied by a constant of 0.03)

Areas

Many different kinds of data are gathered for areas, such as Census divisions, towns, counties, states and other administrative units. For mapping purposes it is important to distinguish between *absolute quantities* which relate to areas and *density measures* or other ratios. Each type of data is associated with particular forms of map.

Absolute quantities: dot mapping. Dot maps are used to show absolute quantities which relate to areas, such as the numbers of persons or the numbers of livestock in a country. Each dot represents a given number of occurrences, for example, one dot for 1000 people or 100 pigs. Dot maps are best suited to phenomena whose distributions are sporadic, because dots are themselves point symbols.

Four important considerations when constructing a dot map are:

  i selecting a suitable dot value;
 ii choosing an appropriate dot size;
iii locating the dots;
 iv drawing the dots.

Selecting a suitable dot value. Consider the range of data to be shown and its distribution. Experiment with dot values. If dot values are too large there will be too few dots; if dot values are too low there will be too many dots and the map will become overcrowded and unclear (Figure 4.28). A rule of thumb is that on small scale maps dot values must be high; on large scale maps dot values can be lowered.

Choosing an appropriate dot size. Dots that are too large will give the map a very coarse appearance and overfill the spaces; dots that are too small will result in a map that looks too light and empty. The dot size and value must be such that the dots just coalesce in areas of greatest density (Figure 4.28).

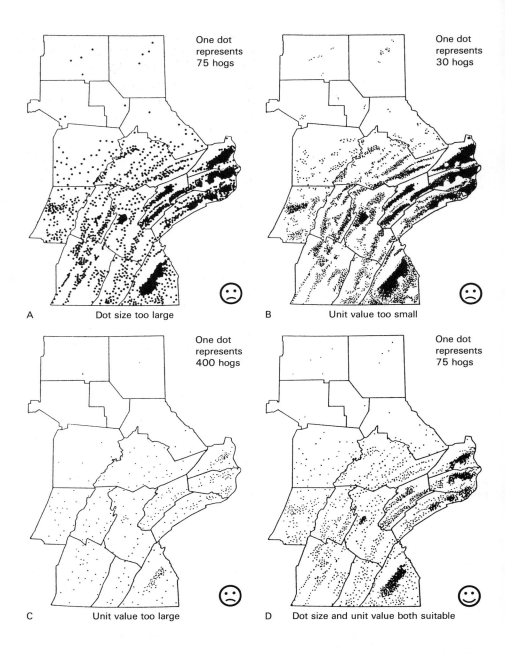

**Figure 4.28
Dot maps: selecting suitable dot
size and value**

A — Dot size too large — One dot represents 75 hogs

B — Unit value too small — One dot represents 30 hogs

C — Unit value too large — One dot represents 400 hogs

D — Dot size and unit value both suitable — One dot represents 75 hogs

**Figure 4.29
Dot maps: locating the dots for a
population distribution**

Locating the dots. The first step in locating the dots is to investigate those factors that control the distribution to be mapped. This may include information on topography or land use. By a process of *progressive elimination* of areas where a distribution is *unlikely* to occur a mapping surface is generated. Dots can be located on this mapping surface with the intention of creating an impression of the actual distribution (Figure 4.29).

I) Dots spread throughout an agricultural area

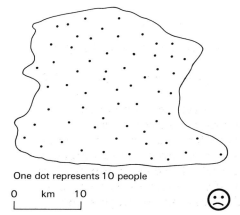

One dot represents 10 people

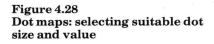
0   km   10

II) Information on topography, soils and settlement

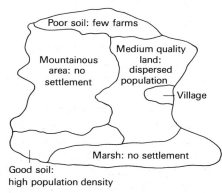

Poor soil: few farms

Mountainous area: no settlement

Medium quality land: dispersed population

Village

Good soil: high population density

Marsh: no settlement

III) Dots located by progressive elimination

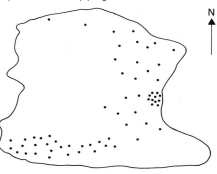

N

84

Where no other regional information is available dots can be spread evenly throughout the area. Try to avoid placing dots in parallel rows or in lines parallel to boundaries as this gives an effect of a regular distribution.

Drawing the dots. Ensure that all dots are of uniform size and shape. Stencils and templates are available for this purpose. When drawing the dots the pen-nib should be kept clean or the dots will become larger and less uniform in shape as the work progresses.

Relative merits. The dot map provides a clear visual impression of a distribution. It is a relatively simple process to count the number of dots in order to estimate absolute figures. The choice of a suitable value and size of dot is important and many problems can arise as a result of inappropriate selection.

Variations of dot maps. Occasionally, it may be useful to employ two sizes of dots for special purposes. For example, in order to accommodate an extreme range of data it may be necessary to assign different values and sizes to dots, say, 1 large dot for 1000 persons and 1 small dot for 100 persons. When two distributions need to be shown on the same map different coloured dots can be used.

Density measures and other ratios: choropleth mapping. Choropleth or shading maps are used to portray *relative data* which relate to area; the prefix *choros* is derived from a Greek root and means *place*. The quantities must be expressed as *densities per units of area,* or as some other amount independent of area like *ratios, percentages* or *per capita statistics.*

There are three stages in the preparation of a choropleth map:

i deriving the values to be mapped;
ii grouping the data;
iii selecting a shading sequence.

Deriving the values to be mapped. Choropleth mapping takes into account the different sizes of administrative regions and presents data in a relative form. Administrative regions are rarely of uniform size. If absolute values are used, large and small areas with similar populations would be grouped together. In the case of a choropleth map, conversion of these data into relative values, such as density per unit area, highlights the sparser distribution of the large area. Densities are calculated in the following way:

$$\frac{\text{Population}}{\text{Area}} = \text{Density}$$

Grouping the data. Once the data are in the right form the next stage is to group the data into *classes*. This is carried out partly because it is difficult to create shading sequences consistent with each unique data value. When selecting class intervals it is important to consider:

i the purpose of the map;
ii the range of data;
iii the spread of data within a distribution. If some data are truly exceptional they should appear so.

Four *intervals* are commonly used to sort data into classes:

i an *arithmetic interval*, whereby values change in a consistent sequence (0–9, 10–19, 20–29, 30–39);
ii a *geometric interval*, whereby values change by a regular ratio (0–4, 5–14, 15–44, 45–134);
iii an *irregular interval*, whereby the data are grouped by means of a *dispersion diagram* (Figure 4.30). A dispersion diagram is a simple device consisting of a vertical scale graduated according to the range of values to be shown. Wherever a particular value occurs a dot is placed on the appropriate point of the vertical scale. The dots usually fall into a pattern of clusters with marked gaps. The *natural breaks* represent the division between groups. The goals always are to recognise natural

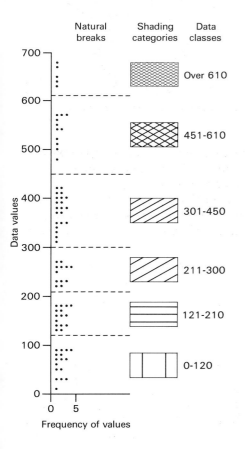

Figure 4.30
Dispersion diagram

I) Examples of hand drawn graded shading

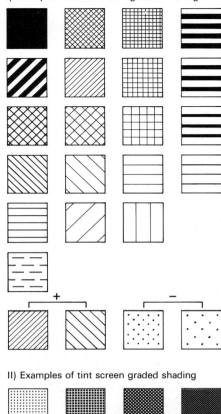

II) Examples of tint screen graded shading

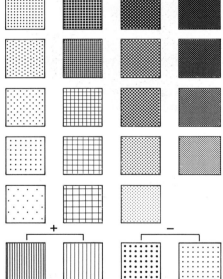

Figure 4.31
**Hand drawn and tint screen
shading**

breaks in the distribution; to minimise differences within groups; to
maximise differences between groups;

iv a *sectional interval*, whereby the data on the dispersion diagram are
divided into groups of equal number: quartiles (four sections); quintiles
(five sections); sextiles (six sections).

Generally, no more than six classes are necessary; more than this makes
visual assessment difficult. It is important that classes are mutually
exclusive and that no overlaps occur at the end of one class and the
beginning of another. For example, a class interval of 0–5, 5–10, 10–15 is
ambigious as the values 5 and 10 fall into two classes; instead the classes
need to be organised into discrete groups, 0–4.9, 5–9.9, 10–14.9, or 0–4, 5–
9, 10–15.

Selecting a shading sequence. A suitable shading system is chosen to
represent the groupings. Two principles should be followed:

i shadings need to be graduated from light to dark. Areas of highest
value or greatest intensity should be given the darkest tones, areas of
lowest value should be given the lightest tones;
ii each shading must have sufficient individuality to be readily
identifiable in the series.

The impression of lighter or darker tones can be created by hand-drawn
effects or by applying ready-made tint screens (Figure 4.31). Ideally,
patterns should be held constant and only tone varied. Sequences which
mix dot and line patterns should be avoided. However, for data that
include positive and negative values, or net loss versus net gain, then two
different shading patterns can be used.

In a hand-drawn series of graded shadings individuality depends on four
factors (Figure 4.31):

i direction or angle at which lines are drawn;
ii distance between the lines;
iii thickness or weight of lines;
iv form of the lines, whether they are solid, broken or dotted.

Tonal differences are usually effected by drawing the lines closer together
and/or by increasing the thickness of the line. At the higher end of the
scale cross-hatching may be added. Black and white can be used at the
extremities of a range, although white is better reserved for a complete
absence of data.

A convenient method of drawing line shadings is to superimpose a
tracing over a sheet of graph paper which has lines the desired width
apart, and ink over the lines on the tracing paper. Ready-made tint screens
include Letratone and Mecanorma. These come in sheets which can be
overlaid and cut to fit the mapped boundaries. Although simple to use they
are relatively expensive.

All shading categories need to be shown and labelled in the key.

Relative merits. The choropleth method is useful for mapping a great
variety of geographical data, such as population, agriculture and other
economic statistics which relate quantity to area. Its advantages are that
the map is simple to construct and it gives a clear visual impression of the
data. The main disadvantages are that it gives the impression of
uniformity of distribution within each area and it emphasises the
boundaries between administrative areas. Obviously, the larger the size of
the areal unit chosen the more generalised is the resulting map.

Variations of the choropleth map: dasymetric mapping. Dasymetric
mapping is a refinement of choropleth mapping. It depends on the
geographical interpretation of area data and the use of local knowledge.
For example, if a map is to be drawn of population density and it is known
that in parts of a region there is no settlement because of inhospitable
terrain, these areas can be excluded and the population density
recalculated for the populated area (Figure 4.33). Dasymetric adjustment
of this kind produces a more accurate and meaningful map.

N

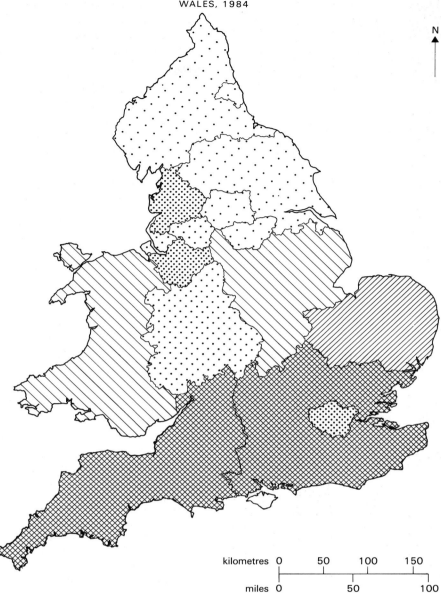

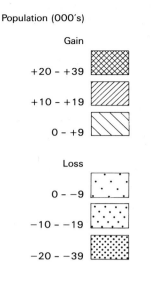

Population (000's)

Gain

+20 – +39

+10 – +19

0 – +9

Loss

0 – –9

–10 – –19

–20 – –39

Figure 4.32
Choropleth mapping

| kilometres | 0 | 50 | 100 | 150 |
| miles | 0 | | 50 | | 100 |

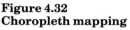

I) Unadjusted population density

Population density

100/km²

50/km²

Area = 100km²
Population = 5 000
Density = 50/km²

II) Dasymetric adjustment to population density

Population density

100/km²

50/km²

Area = 50km²
Population = 5 000
Density = 100/km²

Mountain: no settlement

Figure 4.33
Dasymetric mapping

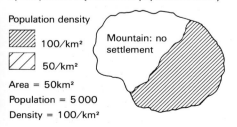

Lines

Quantitative data that occur in lines essentially relate to flows of some kind. These include migrations of people, traffic flows, shipments of goods, movements of materials like crude oil and the volume of telephone calls. All of these flows can be symbolised by a flow-line map.

Flow-line mapping. Two important stages in the preparation of a flow-line map are:

  i selecting a linear scale;
 ii depicting the direction of flow.

Selecting a linear scale. Flow-line maps show the volume of movement or the amount of goods carried along any routeway by *varying the width of the line in proportion to the flow.* In order to arrive at a suitable linear scale consider the range of data to be plotted. Choose a scale of line thickness that can be fitted into the map without confusion. For example, 1mm of line width may represent 5 trains per day; other values would then be represented by lines of proportional width to the nearest 5 trains. Routes with less than 5 trains can be represented by a dotted line. Where the range of values is large a convenient modification is to group the data into classes and to adopt a series of graduated band-widths for each of the intervals. A method of grouping data is shown in Figure 4.34. Although the lines will not be strictly proportional the map will distinguish the major lines of flow from those of lesser importance.

Depicting the direction of flow. The quantities shown by flow-line maps are either:

i flows along routeways of various kinds, in which case the method is to draw astride each route a band of proportional width (Figure 4.34); or
ii dynamic data not strictly tied to a route, such as migrations of people to different countries. These lines are shown by drawing a line of proportional thickness from origin to destination (Figure 4.36).

Figure 4.34
Flow-line mapping: flows along known routes

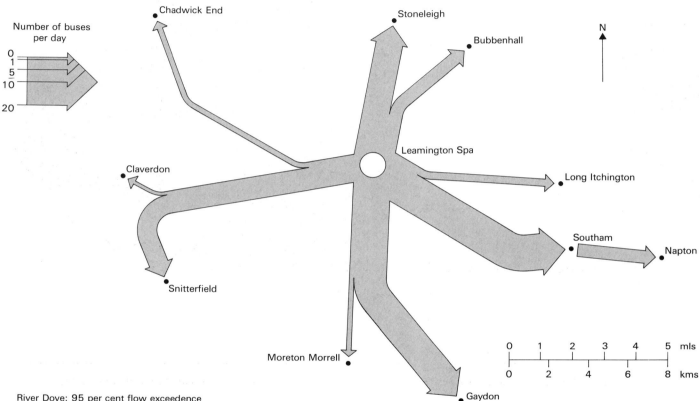

BUS SERVICES FROM LEAMINGTON SPA TO A SELECTION OF VILLAGES IN SOUTH WARWICKSHIRE, 1985

In both cases arrow heads are used to show the direction of this movement. Flow-lines may be filled-in in solid black, tinted or left open. Different methods of keying flow-lines are shown in Figures 4.34 and 4.36.

Relative merits. Flow-line mapping is relatively straightforward and gives an immediate visual impression of comparative volumes of movement. Problems include:

i selecting a suitable scale of band-width, especially if values are very widely spread;
ii individual flow-lines may branch, subdivide or converge and in drawing these some attempt should be made to maintain the appearance of continuity (Figure 4.34).

Variations of flow-line maps:

i *Desire lines* are a particular variation of flow-line mapping. These simply symbolise linkage, each line joins the place of origin and destination of a particular movement. Desire lines are usually used to show such movements as where people travel for particular services or where people go to work (Figure 4.37);
ii *Network line* diagrams. To meet the need for a convenient representation of a network and its flows, places can be shown in terms of their position on the network, not in terms of their location on a

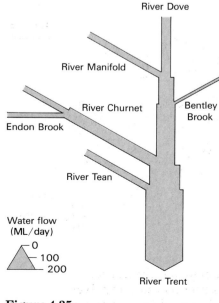

River Dove: 95 per cent flow exceedence megalitres per day (ML/day)

Figure 4.35
Network line diagram

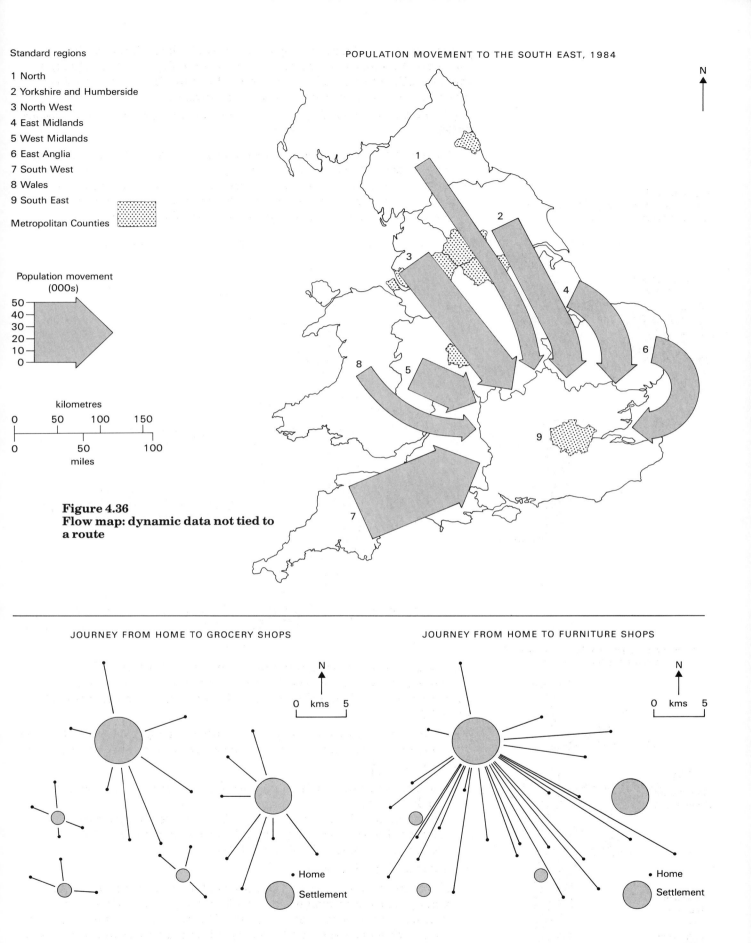

Standard regions

1 North
2 Yorkshire and Humberside
3 North West
4 East Midlands
5 West Midlands
6 East Anglia
7 South West
8 Wales
9 South East

Metropolitan Counties

Population movement
(000s)

50
40
30
20
10
0

kilometres

0   50   100   150

0   50   100
         miles

Figure 4.36
**Flow map: dynamic data not tied to
a route**

N

JOURNEY FROM HOME TO GROCERY SHOPS

N

0   kms   5

JOURNEY FROM HOME TO FURNITURE SHOPS

N

0   kms   5

• Home

Settlement

• Home

Settlement

Figure 4.37
Desire line maps

topographic map. An example of this type of map is the London
Underground. In hydrological studies it is useful to simplify river
systems into network line diagrams. In this way different flows along
the river system can be shown (Figure 4.35).

EXERCISE 1: ISOLINE MAPPING

Figure 4.38 shows part of a valley and river system which has recently been surveyed. All spot heights are in metres.

Q **1** Trace an outline of the area and draw a contour map. Interpolate contours at 20m vertical intervals. Start with a 50m contour line.

2 Complete the map by providing full marginal information. The scale of the map is 1:10 000.

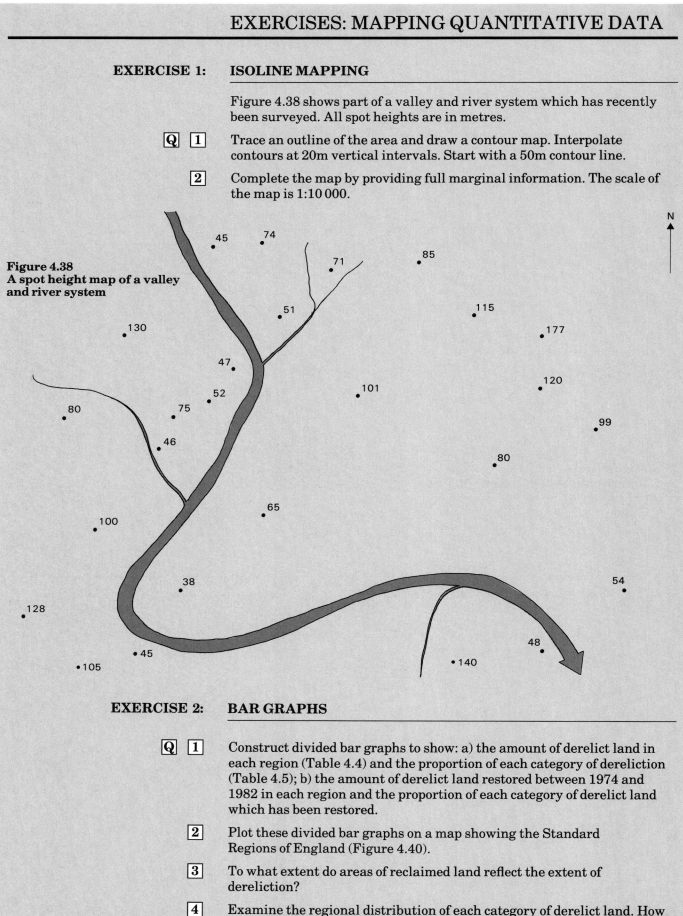

**Figure 4.38
A spot height map of a valley and river system**

EXERCISE 2: BAR GRAPHS

Q **1** Construct divided bar graphs to show: a) the amount of derelict land in each region (Table 4.4) and the proportion of each category of dereliction (Table 4.5); b) the amount of derelict land restored between 1974 and 1982 in each region and the proportion of each category of derelict land which has been restored.

2 Plot these divided bar graphs on a map showing the Standard Regions of England (Figure 4.40).

3 To what extent do areas of reclaimed land reflect the extent of dereliction?

4 Examine the regional distribution of each category of derelict land. How successful has each region been in reclaiming these different types of dereliction?

TABLE 4.4 Derelict land: by category of dereliction: by region: at 1 April 1982

Standard Regions	Spoil heaps	Excavation and pits	Military dereliction	Derelict railway land	Other forms of dereliction	All derelict land	All derelict land as at 1 April 1974
HECTARES							
North	1872	1043	168	1375	2849	7307	9411
Yorkshire & Humberside	1070	1433	385	1428	1115	5431	5451
East Midlands	1225	1258	644	1339	732	5198	5171
East Anglia	15	305	251	170	63	804	1783
Greater London (GL)	45	382	364	181	982	1954	324
South East (excl GL)	57	1439	268	374	387	2525	2036
South West	4870	420	208	820	317	6635	6415
West Midlands	2174	917	330	875	1491	5787	4667
North West	2012	1381	398	1648	4603	10042	8015
England	13340	8578	3016	8210	12539	45683	43273

TABLE 4.5 Derelict land restored, 1 April 1974 to 31 March 1982: by category of dereliction: by region

Standard Regions	Spoil heaps	Excavation and pits	Military dereliction	Derelict railway land	Other forms of dereliction	All derelict land
HECTARES						
North	1431	319	55	391	1496	3692
Yorkshire & Humberside	1229	469	35	245	238	2216
East Midlands	1011	230	224	283	174	1922
East Anglia	–	47	57	186	20	310
Greater London (GL)	25	135	48	37	176	421
South East (excl GL)	3	246	36	18	113	416
South West	1103	139	52	73	66	1433
West Midlands	1385	992	213	335	820	3745
North West	374	459	317	423	1376	2949
England	6124	2306	1547	1996	4479	16452

SOURCE: Survey of Derelict Land in 1982, Department of Environment

EXERCISE 3: PROPORTIONAL CIRCLES

Table 4.6 provides information on the numbers of people classified as living in urban and rural places in selected European countries in 1986.

Q **1** On a tracing paper overlay of Figure 4.39 construct proportional circles to demonstrate the total populations of each country and divide the circles to illustrate the percentage of each population classified as urban and rural.

2 Describe and account for the variations in the spatial distribution of urban and rural populations.

3 What are the problems associated with defining rural areas and their populations?

TABLE 4.6
Population classified as urban and rural, 1986

	Urban	Rural	Total
Austria	4161045	3394293	7555338
Denmark	4297092	826897	5123989
Finland	2903807	1951979	4855786
France	39799820	14473380	54273200
Eire	1914785	1528620	3443405
Netherlands	7360960	7032339	14393299
Norway	2893193	1197939	4091132
Portugal	2918549	6914465	9833014
Spain	23555642	13143528	36699170
Sweden	6913493	1406945	8320438
Switzerland	3632590	2733370	6365960
UK – (Scotland)	4474096	559846	5033942

SOURCE: United Nations Demographic Yearbook, 1986

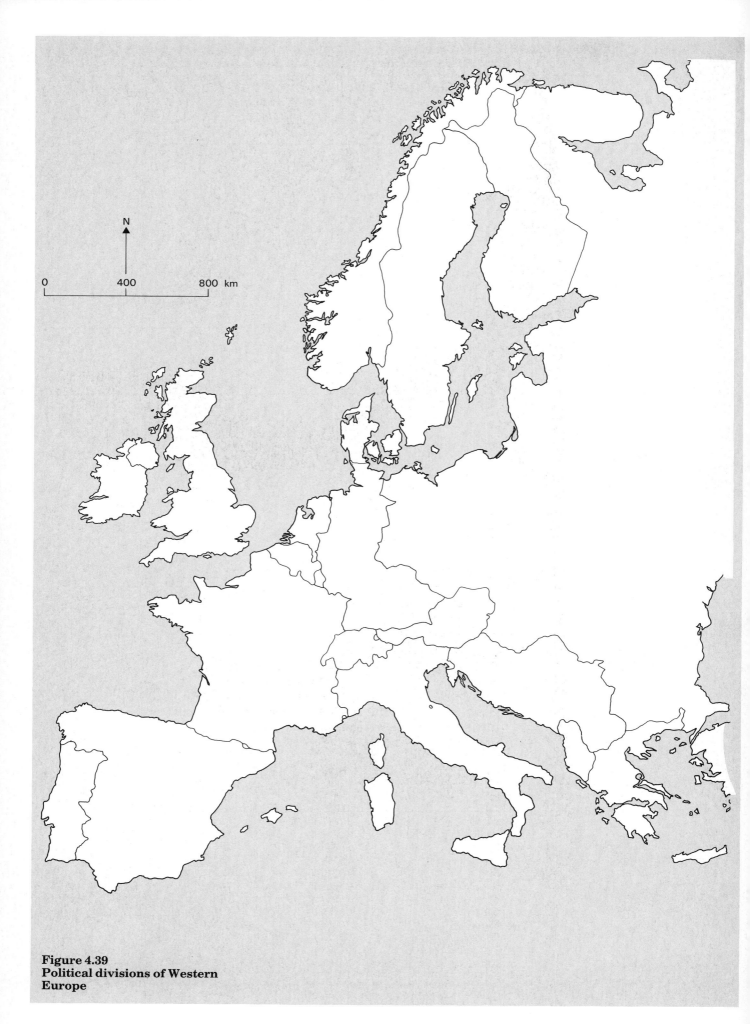

**Figure 4.39
Political divisions of Western
Europe**

N

| 0 | 400 | 800 km |

Counties

Metropolitan counties and Greater London

1 Tyne and Wear
2 Merseyside
3 Greater Manchester
4 West Yorkshire
5 South Yorkshire
6 West Midlands
7 Greater London

Standard Regions

Northumberland

NORTH

Durham

Cleveland

Cumbria

North Yorkshire

YORKS AND HUMBER

Lancashire

NORTH WEST

Humberside

Cheshire

Derby

Nottingham

Lincolnshire

Clwyd

Gwynedd

Stafford

EAST MIDLANDS

Shropshire

Leicester

Norfolk

EAST ANGLIA

WALES

Powys

Cambridge

Suffolk

WEST MIDLANDS

Warwick

Northampton

Bedford

Dyfed

Hereford and Worcester

Bucking ham

Essex

Gloucester

Oxford

Hertford

Gwent

West Glamorgan

Mid Glamorgan

South

Avon

Wiltshire

Berkshire

SOUTH EAST

Surrey

Kent

Somerset

Hampshire

West Sussex

East Sussex

SOUTH WEST

Dorset

Devon

Cornwall

Figure 4.40
Standard Regions and counties of England and Wales

EXERCISE 4:   PROPORTIONAL CUBES

Table 4.7 shows the GNP/capita in US dollars for selected countries.

Q 1   Construct either proportional cubes or spheres to represent the data. Plot the information on a world map (Figure 4.27).

2   Consider the problems associated with mapping this information by either technique.

TABLE 4.7   GNP/capita in US dollars, 1984

Ethiopa	110	Indonesia	540	UK	8570
Bangladesh	130	Papua New Guinea	710	France	9760
Mali	140	Colombia	1390	Japan	10630
Malawi	180	Chile	1700	Australia	11740
India	260	Brazil	1720	USA	15390
China	310	Mexico	2040	Switzerland	16630
Sri Lanka	360	South Africa	2340		

SOURCE: World Bank, 1986

EXERCISE 5:   DOT MAPPING

Table 4.8 provides data on the number of pig and poultry farms in English counties in 1985.

Q 1   Trace an outline of the English counties shown in Figure 4.40. Draw a dot map to show the distribution of pig and poultry farming. When devising your map, experiment with dots of equal value and size and of different value and size.

2   Justify your method of dot mapping.

3   Describe the distribution of pig and poultry farms in England. Which counties are associated with high numbers and which counties have low numbers?

4   Suggest factors to account for this distribution.

5   What other characteristics of pig and poultry farms need to be known in order to gain a better understanding of their overall distribution?

TABLE 4.8
Pig and poultry farms in English counties, 1985

Counties	No. of farms	Counties	No. of farms
Avon	58	Merseyside	23
Bedford	49	Greater London	24
Berkshire	45	Norfolk	442
Buckingham	88	Northampton	70
Cleveland	21	Tyne and Wear	11
Cambridge	145	Northumberland	30
Cheshire	142	Nottingham	114
Cornwall	69	Oxfordshire	107
Cumbria	65	Salop	111
Derby	75	Somerset	145
Devon	209	Stafford	99
Dorset	96	Suffolk	407
Durham	50	Surrey	59
Essex	216	Sussex, East	60
Gloucester	74	Sussex, West	58
Hampshire	141	Warwick	69
Isle of Wight	12	Greater Manchester	66
Hereford/Worcester	172	Wiltshire	128
Hertford	51	West Midlands	29
Kent	125	South Yorkshire	66
Lancashire	256	North Yorkshire	425
Leicester	83	West Yorkshire	147
Lincoln	249	Humberside	445
		Total	5626

SOURCE: MAFF, 1986

EXERCISE 6: CHOROPLETH MAPPING

Q 1 Trace an outline of the counties of England and Wales shown in Figure 4.40. Draw a choropleth map to represent rates of unemployment in English and Welsh counties (Table 4.9).

2 Does the county pattern of unemployment support the notion of a North-South divide? Are there any regional exceptions to an overall trend?

3 Suggest and justify *five* other measures which could be used to consider broad patterns of social and economic inequality.

TABLE 4.9   Regional and county unemployment rates in England and Wales, October 1985

Region	County	Unemployment rate (%)	Region	County	Unemployment rate (%)	Region	County	Unemployment rate (%)
North		18.9	*South West*		12.1	*South East*		9·9
	Cleveland	22·4		Avon	11·4		Bedfordshire	10·5
	Cumbria	12·5		Cornwall	18·2		Berkshire	7·3
	Durham	18·7		Devon	13·8		Buckinghamshire	8·2
	Northumberland	15·6		Dorset	11·4		East Sussex	11·9
	Tyne and Wear	20·3		Gloucestershire	10·1		Essex	12·1
Yorkshire and Humberside		15.1		Somerset	10·8		Greater London	10·7
	Humberside	16·6		Wiltshire	9·9		Hampshire	9·9
	North Yorkshire	10·7					Hertfordshire	7·2
	South Yorkshire	17·9	*West Midlands*		15.6		Isle of Wight	14·9
	West Yorkshire	14·1		Hereford and Worcestershire	13·8		Oxfordshire	7·8
East Midlands		12.5		Shropshire	16·4		Surrey	No data
	Derbyshire	13·6		Staffordshire	14·0		West Sussex	7·1
	Leicestershire	10·5		Warwickshire	12·5	*Wales*		17.2
	Lincolnshire	13·4		West Midlands Metropolitan County	16·7		Clwyd	18·6
	Northamptonshire	11·7					Dyfed	18·2
	Nottinghamshire	13·2	*North West*		16.3		Gwent	17·2
East Anglia		10.5		Cheshire	13·6		Gwynedd	18·9
	Cambridgeshire	9·4		Greater Manchester	15·5		Mid Glamorgan	19·0
	Norfolk	12·4		Lancashire	14·0		Powys	13·1
	Suffolk	9·5		Merseyside	21·1		South Glamorgan	14·4
							West Glamorgan	16·2

SOURCE: Regional Trends, 1986

EXERCISE 7: FLOW-LINE MAPPING

Q 1 Trace an outline of Figure 4.39 and plot a flow-line map to show the number of visits made by UK residents to other parts of Europe (Table 4.10).

2 Table 4.10 shows a wide range of data. What problems are experienced when attempting to show such a range by a flow-line technique?

3 Suggest reasons for the wide variation in the number of visits made to different European countries.

**TABLE 4.10
Number of visits to other parts of Europe by UK residents, 1984**

	Number of visits (000s)		Number of visits (000s)
Belgium/Luxembourg	776	Yugoslavia	477
France	4482	Spain	5022
West Germany	1294	Portugal	573
Italy	1184	Austria	609
Netherlands	868	Switzerland	519
Denmark	136	Norway	139
Eire	1552	Sweden	135
Greece	1048	Finland	38
		Gibraltar/Malta/Cyprus	475

SOURCE: Overseas Travel and Tourism, 1986

Not all geographical data need be mapped. Graphs and diagrams can be used to show:

i absolute changes and rates of change in values over time and through space;

ii simple relationships between two or more quantities;

iii components of quantities.

Variations in quantities over time and through space

Line graphs, whether simple or multiple, may be used to show variations in absolute or percentage values. Graphs consist of two axes, the horizontal or x axis and the vertical or y axis. It is usual to place the *independent* variable along the x axis and the *dependent* variable along the y axis. In order to decide which variable is *independent,* consider which one of the pair of variables to be plotted causes the other to change. For example, rainfall is dependent on the months of the year and not the other way around. In this case, months of the year is the independent variable. Each axis should be clearly labelled with the names of the variables. The scales for each axis should be carefully chosen as these will determine the visual impression given by the graph (Figure 4.41).

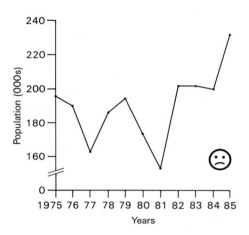

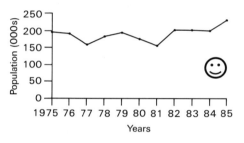

Figure 4.41
Line graphs: the effect of scale on the visual presentation of data: both graphs show the same data

Multiple distributions can be combined on a single graph. For example, rainfall for different countries over the same time period, or the proportion of households with selected consumer durables over a given time period can both be shown in this way. Lines of different colour, thickness or pattern add clarity to the presentation (Figure 4.42).

Bar graphs or columnar diagrams provide a straightforward means of comparing quantities. The widths of each bar are kept constant and their lengths are graduated according to a suitable scale. The bars can be drawn vertically or horizontally (Figure 4.43).

Circular graphs are used for showing a variable which is continuous over time, such as temperature or the number of sunshine hours. There are two axes, the circumference of the circle and the radius. The circumference is normally time; for example, each month of the year is represented by 30° of circumference (360° ÷ 12). Values increase radially outwards (Figure 4.44).

Radial graphs or ray diagrams are those in which lines of proportional length radiate outwards from a central point in a given direction (Figure 4.45). Examples include the wind-rose, which shows the number of hours or percentage of time in which a wind blows from a certain direction. In economic mapping, lines representing export values can be drawn of proportional length in the direction of movement.

Graphs plotted on *arithmetic paper* emphasise absolute changes in values. On other occasions it may be useful to show *rates of change*. This can be done by plotting the data on semi-logarithmic graph paper (logarithmic along the vertical axis only). In this case, the numbers on the vertical scale are not evenly spaced; the interval is governed by the difference between their logarithms and not by the difference between the numbers themselves. There is no need to convert any values into

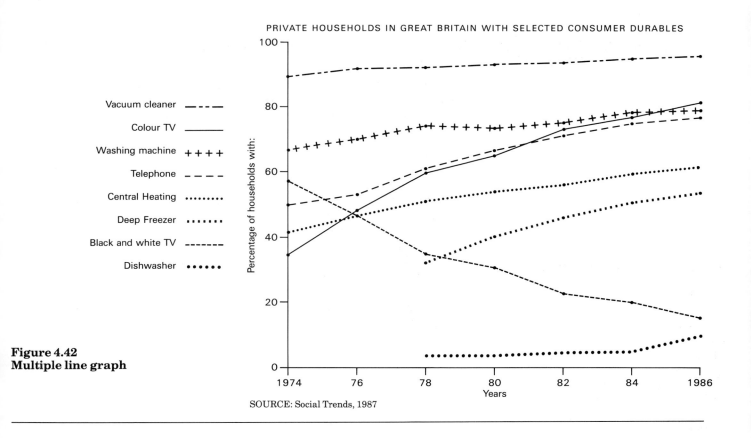

PRIVATE HOUSEHOLDS IN GREAT BRITAIN WITH SELECTED CONSUMER DURABLES

Vacuum cleaner — · — ·

Colour TV ————

Washing machine + + + +

Telephone — — —

Central Heating · · · · · · ·

Deep Freezer · · · · · ·

Black and white TV - - - - - -

Dishwasher · · · · · ·

Figure 4.42
Multiple line graph

SOURCE: Social Trends, 1987

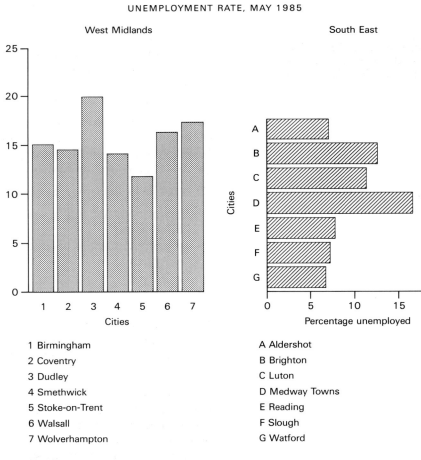

I) Vertical bar charts   II) Horizontal bar charts

UNEMPLOYMENT RATE, MAY 1985

West Midlands   South East

1 Birmingham   A Aldershot

2 Coventry   B Brighton

3 Dudley   C Luton

4 Smethwick   D Medway Towns

5 Stoke-on-Trent   E Reading

6 Walsall   F Slough

7 Wolverhampton   G Watford

SOURCE: Regional Trends, 1986

Figure 4.43
Bar graphs: vertical and horizontal
presentation of data

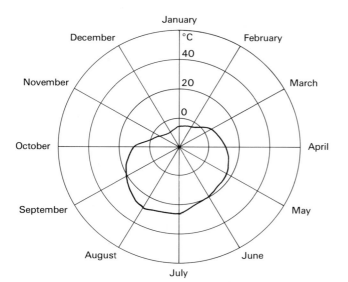

**Figure 4.44
Circular graph**

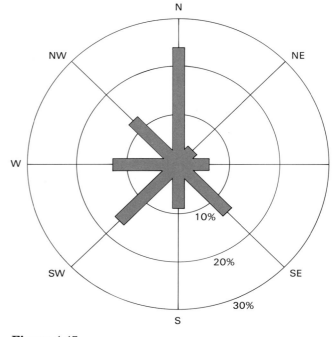

**Figure 4.45
Radial graph: a wind rose**

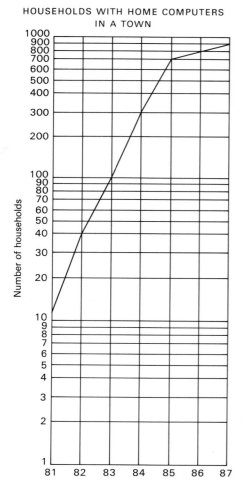

**Figure 4.46
Semi-logarithmic graph: 3-cycle
paper**

logarithms before plotting, simply plot the data onto the logarithmic scale of the paper. Semi-logarithmic graph paper is available in various cycles, one to five, each cycle representing a set of lines (Figure 4.46). The bottom line of each cycle is either 10 or a multiple or decimal of 10. For example, one-cycle graph paper could be used for numbers 1–10, two-cycle for numbers 1–100, three-cycle for numbers 1–1000 and so on up to 100 000 at the top of the five-cycle paper. Before plotting on semi-logarithmic paper the range of values should be examined so that the correct style of paper is used. Another advantage of using semi-logarithmic paper is that a greater range of values can be accommodated than on normal arithmetic paper.

Simple relationships between two or more variables The *scattergram* is the most useful way of examining relationships between two variables. This graph is frequently used as the first stage of more detailed statistical analyses. Its character and form are explained in Chapter 5.

Triangular graphs or ternary diagrams are useful for two reasons;

i they provide a visual assessment of the relationship between three phenomena;
ii they offer a method of classifying data according to the dominance of each phenomenon.

Triangular graphs can only be used if the values of the three phenomena, when taken together, add up to a whole, that is, 100%. In many cases data needs to be converted into percentages and then plotted. The diagram consists of an equilateral triangle, the sides of which form three scales, graduated from 0 to 100% (Figure 4.47). Provided that the three scores total 100% they can be shown by a single point on the diagram.

Triangular graphs can be used to show any data which can be conveniently divided into three parts, such as the relative proportion of clay, silt and sands in soils or the proportion of households in owner-occupation, private renting and council tenancy in different areas of a city.

The position of any point within the triangle reflects the relative dominance of each phenomenon. Figure 4.48 shows how data may be classified according to its plotted position.

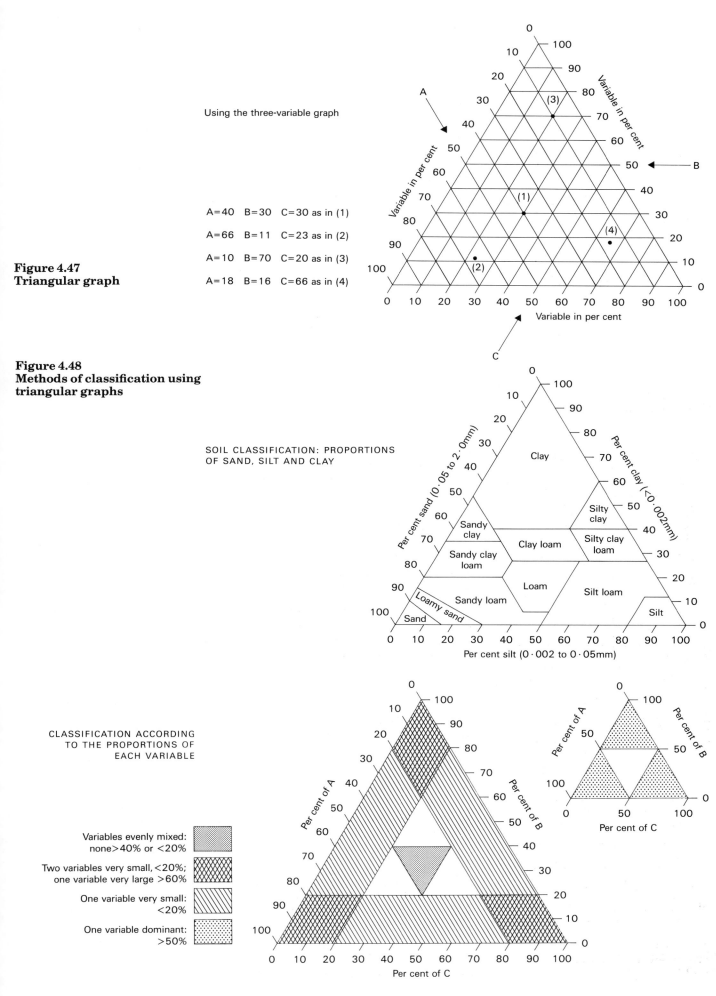

Using the three-variable graph

A=40 B=30 C=30 as in (1)

A=66 B=11 C=23 as in (2)

A=10 B=70 C=20 as in (3)

A=18 B=16 C=66 as in (4)

**Figure 4.47
Triangular graph**

**Figure 4.48
Methods of classification using
triangular graphs**

SOIL CLASSIFICATION: PROPORTIONS
OF SAND, SILT AND CLAY

CLASSIFICATION ACCORDING
TO THE PROPORTIONS OF
EACH VARIABLE

Variables evenly mixed:
none>40% or <20%

Two variables very small,<20%;
one variable very large >60%

One variable very small:
<20%

One variable dominant:
>50%

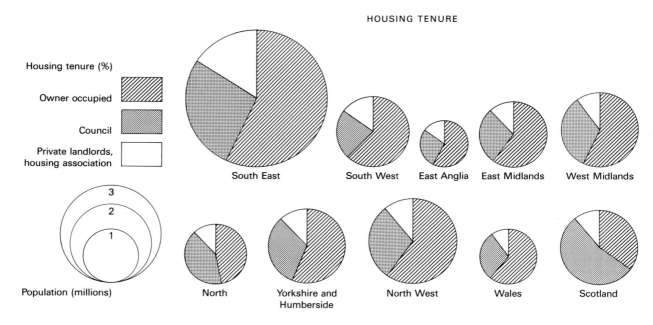

HOUSING TENURE

Housing tenure (%)

Owner occupied

Council

Private landlords,
housing association

Population (millions)

South East

South West

East Anglia

East Midlands

West Midlands

North

Yorkshire and
Humberside

North West

Wales

Scotland

**Figure 4.49
Divided proportional circles**

Components of quantities *Divided bar graphs* and *divided circles* (pie graphs) are used for showing a quantity which can be divided into components. The length of the bar and the area of the circle (Figure 4.49) are made proportional to the total quantity. It is not essential for these diagrams to be plotted on a map.

In effect, the *population pyramid* is another form of divided bar graph. Population pyramids are used to show the age and sex composition of a population (Figure 4.50). The shape of the pyramid gives a number of clues as to the processes that have gone into producing the particular population

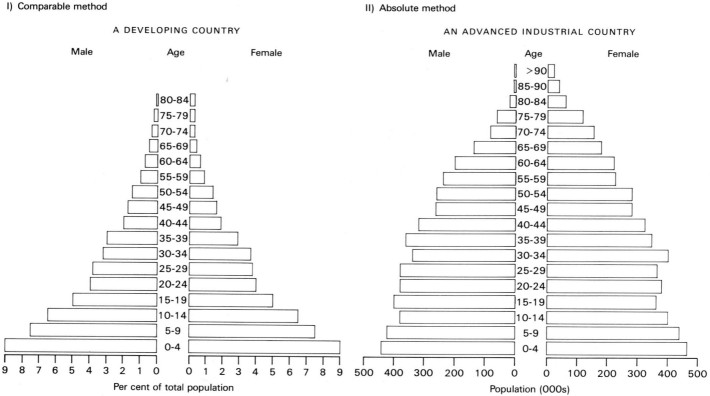

I) Comparable method

A DEVELOPING COUNTRY

Male   Age   Female

Per cent of total population

II) Absolute method

AN ADVANCED INDUSTRIAL COUNTRY

Male   Age   Female

Population (000s)

**Figure 4.50
Population pyramids**

100

structure. Three major steps are involved in the preparation of population pyramids:

i divide the population into age groups; five-year intervals are conventionally used. Five-year divisions not only provide the geographer with sufficient detail to make informed observations about population structure, but also most international returns publish data in this form;

ii choose either the *absolute method* or *comparable method* for showing the population data. In the absolute method total numbers for each age and sex are plotted. The advantage of this technique is that it is simple to carry out, but problems arise when places of greatly varying size are to be compared. In the comparable method each age group by sex is represented as a percentage of the total population for that sex. For example, the number of boys aged 5 to 9 is expressed as a percentage of the total male population. This technique gives an overall impression of population structure and highlights variations in the shape of different pyramids;

iii plot the data on a suitable scale. The vertical axis shows age groups in five-year intervals. The horizontal axis represents the numbers or percentage of people in each of these age groups. Males are plotted to the left and females to the right of the vertical axis.

4.9 SUMMARY

Some of the different types of thematic maps and diagrams discussed in this chapter are shown in Figure 4.51.

Figure 4.51
A summary of different types of thematic maps

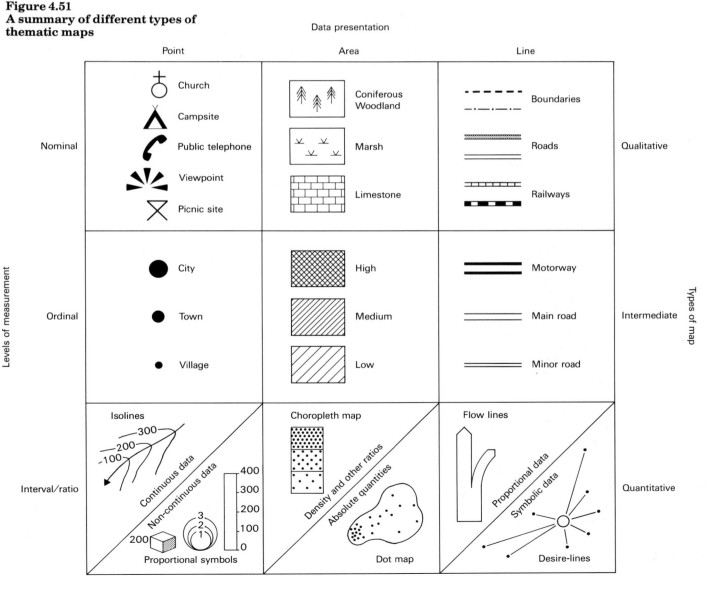

EXERCISES: DRAWING DIAGRAMS AND GRAPHS

EXERCISE 1: MULTIPLE LINE GRAPHS

Q 1 Draw a multiple line graph to show the changes in sulphur dioxide emissions in the UK over the period 1970–1985 (Table 4.11).

2 Describe the trends in sulphur dioxide emission for this period.

3 Suggest factors to account for the observed variations in emissions.

TABLE 4.11
Sulphur dioxide: estimated emissions (tonnes) from fuel combustion by type of consumer in the United Kingdom

Sources of emission from fuel combustion:	'70	'71	'72	'73	'74	'75	'76	'77
Power stations	2·77	2·80	2·87	3·02	2·78	2·82	2·69	2·74
Industry	2·36	2·19	2·02	2·07	1·89	1·74	1·72	1·66
Domestic	0·52	0·46	0·37	0·36	0·35	0·30	0·28	0·29
All sources	6·12	5·86	5·65	5·81	5·36	5·17	5·00	5·00

Sources of emission from fuel combustion:	'78	'79	'80	'81	'82	'83	'84	'85
Power stations	2·81	3·01	2·87	2·71	2·62	2·53	2·50	2·53
Industry	1·66	1·68	1·31	1·05	0·95	0·77	0·67	0·67
Domestic	0·26	0·26	0·22	0·21	0·20	0·20	0·16	0·20
All sources	5·02	5·34	4·67	4·22	4·01	3·69	3·54	3·58

SOURCE: Warren Spring Laboratory, 1986

EXERCISE 2: BAR GRAPHS

Q 1 Construct bar graphs to show variations in river water quality by Water Authority area for the period 1978 – 1985 (Table 4.12).

2 Suggest ways in which poor quality water can be recognised.

3 Account for the higher proportion of river length classed as either of bad or of poor quality in the North West, Yorkshire and Anglian Water Authority areas compared to elsewhere.

4 Why should water quality vary from year to year?

TABLE 4.12
River water quality: proportion of polluted river length by Water Authority area, 1978 to 1985

Water Authority area	Percentage of all classified river length classed as either bad or of poor quality						
	1978/79	1979/80	1980/81	1981/82	1982/83	1983/84	1984/85
North West	15·3	15·3	15·5	15·0	15·1	19·0	19·5
Yorkshire	13·5	13·25	13·0	12·5	12·0	12·8	13·3
Anglian	13·7	9·8	12·0	11·5	10·0	10·5	9·9
Southern	6·0	4·9	4·8	4·0	3·5	5·5	4·0
Wessex	5·0	5·0	5·5	5·0	4·9	4·9	4·8
South West	5·1	5·3	5·2	4·9	6·0	5·0	6·2

SOURCE: Water Authorities Association, 1986

EXERCISE 3: DIVIDED BAR GRAPH

Q 1 Using the data in Table 4.13, draw a divided bar graph to show both the total amount and the composition of timber production for the period 1977–2001.

2 Suggest reasons why coniferous woodland is expected to increase by the year 2001.

3 Why did broadleaf woodland decline until 1982 and stabilise from this period onwards?

		Coniferous woodland			
TABLE 4.13 Past and future timber production in Great Britain $\left(\begin{array}{c} m^3 \text{ overbark} \\ \times 1000 \end{array}\right)$	Year	Forestry Commission	Private woodland	Broadleaf woodland	Total
	1977	1890	740	1450	4080
	1978	1950	760	1400	4110
	1979	2060	880	1400	4340
	1980	2410	890	1300	4600
	1981	2530	730	1300	4560
	1982	2690	860	1000	4550
	1983	2770	940	900	4610
	1984	2760	1110	900	4770
	1985–86	2900	1600	900	5400
	1987–91	3400	2000	900	6300
	1992–96	4500	2500	900	7900
	1997–2001	5600	3200	900	9700

[1]Almost all broadleaved production is derived from private land

SOURCE: Forestry Commission, Forestry Facts and Figures 1984–85

EXERCISE 4: PROPORTIONAL CIRCLES AND PIE CHARTS

Table 4.14 shows the population in private households by the Standard Regions of Britain, classified by ethnic origin.

Q **1** Construct proportional circles to show the total numbers of people classified as *other than white* for each Standard Region, and divide each circle to show the percentage of these different ethnic groups.

2 Compare the regional distribution of different ethnic groups within Britain.

3 Why should some regions, such as the North, have relatively small numbers of non-whites compared to regions like Greater London, the South East and the West Midlands?

TABLE 4.14 Population (000s) in private households, classified by ethnic origin, 1984

Standard Region	White	West Indian	Indian	Pakistani	Bangladeshi	Chinese	African	Other	Total (excluding Whites)
North	2974	–	5	5	–	10	4	61	85
Yorkshire and Humberside	4578	28	41	94	7	6	3	91	270
West Yorkshire	1859	22	38	74	6	2	2	33	177
East Midlands	3647	24	72	12	–	4	2	70	184
South East	15242	362	412	74	68	59	84	502	1561
Greater London	5497	307	318	43	60	37	71	105	922
South West	4262	8	9	5	–	5	3	76	106
West Midlands	4689	72	161	94	11	5	1	86	430
West Midlands Metropolitan County	2264	67	148	84	6	4	1	46	356
North West	5989	28	81	40	6	17	9	145	326
Greater Manchester	2388	21	46	25	6	2	4	74	178
Wales	2706	3	12	8	–	2	1	40	66
Scotland	4950	–	9	38	–	–	–	64	111

SOURCE: Regional Trends, 1986

EXERCISE 5:   CIRCULAR GRAPHS

The mean monthly temperatures for selected places are given in Table 4.15

Q 1   For each place construct a circular graph to show the monthly variation in temperature.

2   Consider the relative merits of using circular graphs as opposed to line graphs to show the same data.

TABLE 4.15
Mean monthly temperatures in degrees centigrade

	J	F	M	A	M	J	J	A	S	O	N	D
Khartoum (Sudan)	22	25	28	30	31	32	29	20	30	31	26	21
Johannesburg (S. Africa)	19	18	17	15	12	10	9	12	16	17	19	20
Tashkent (USSR)	-2	4	9	14	20	25	28	26	20	12	8	3
Oxford (UK)	5	6	8	9	12	14	15	15	13	10	8	5

EXERCISE 6:   RADIAL GRAPHS

Table 4.16 shows joint directions of one of the granite tors of Dartmoor.

Q 1   Draw a radial graph to represent the number of joints in each direction.

2   Look at a 1:50 000 map of Dartmoor and consider whether joint directions affect river valley directions.

TABLE 4.16
Joint directions of Great Staple Tor

Direction	No. of joints	Direction	No. of joints	Direction	No. of joints
0°/360	20	120	63	240	4
10	50	130	20	250	5
20	40	140	47	260	18
30	45	150	30	270	65
40	30	160	25	280	67
50	12	170	13	290	80
60	4	180	21	300	61
70	5	190	52	310	20
80	20	200	41	320	48
90	65	210	45	330	40
100	68	220	33	340	15
110	80	230	10	350	14

EXERCISE 7:   TRIANGULAR GRAPHS

Table 4.17 shows the distribution of Gross Domestic Product derived from agriculture, industry and services for selected countries for 1965 and 1984.

Q 1   For each year plot the data on a separate triangular graph.

2   Use the triangular graphs to classify the countries according to their structures of production.

3   What changes have taken place in the structures of production between 1965 and 1984.

TABLE 4.17
Distribution of Gross Domestic Product (per cent)

	Agriculture		Industry		Services	
	1965	1984	1965	1984	1965	1984
China	39	36	38	44	23	20
Congo People's Republic	29	7	27	60	44	33
Ethiopia	58	48	14	16	28	36
Hong Kong	2	1	40	21	58	78
India	47	35	22	27	31	38
Indonesia	59	26	12	40	29	34
United Kingdom	3	2	41	36	56	62
USA	3	2	38	32	59	66

SOURCE: World Bank, 1986

EXERCISE 8:   POPULATION PYRAMIDS

In Table 4.18 data on the age and sex composition of four different countries is shown.

Q 1 Construct population pyramids a) by the *absolute method* to compare the population structure of Scotland and Rwanda, and b) by the *comparative method* to compare the population structure of Japan and Brazil.

2 Describe the differences in population structure in both cases.

TABLE 4.18
Population composition

Age Group	Scotland Male	Scotland Female	Rwanda Male	Rwanda Female
0–4	164955	157901	460648	464428
5–9	166960	159431	337632	341024
10–14	231399	205039	292609	293833
15–19	235319	227288	284706	285657
20–24	216412	211099	229097	233600
25–29	183807	180642	157790	169795
30–34	177123	174921	99380	118974
35–39	165739	165283	84331	111738
40–44	143942	148384	80458	102092
45–49	140639	147849	70557	83655
50–54	141181	151921	61508	75233
55–59	137378	151211	50429	61086
60–64	122146	146037	44030	52339
65–69	103302	134231	30978	31741
70–74	85391	124423	13984	13273
75–79	53392	96192	23235 }	22291 }
80–84	24506	59906		
85+	11019	38187	9775	8527
Total	2486610	2679947	2331147	2469286

Age Group	Japan Male	Japan Female	Brazil Male	Brazil Female
0–4	3990014	3782833	9007000	8868000
5–9	4740661	4506037	8048000	8011000
10–14	5071070	4816645	7529000	7453000
15–19	4416242	4205842	7307000	7317000
20–24	4043662	3913919	6324000	6299000
25–29	4018665	3961932	5365000	5365000
30–34	5117094	5124915	4342000	4333000
35–39	4835683	4801473	3460000	3445000
40–44	4460792	4481694	3013000	3016000
45–49	4129481	4166932	2521000	2543000
50–54	3819217	3884881	2197000	2230000
55–59	3153820	3422119	1769000	1823000
60–64	2082299	2803892	1342000	1401000
65–69	1764392	2306352	1057000	1129000
70–74	1443529	1946387		
75–79	922571	1303539	1432000 }	1628000 }
80–84	501313	813602		
85+	219059	451014		
Total	58789564	60693925	64793000	64861000

SOURCE: United Nations Demographic Yearbook, 1986

Chapter 5 DATA DESCRIPTION AND ANALYSIS

5.1 INTRODUCTION

In Chapter 4, you were shown how to construct maps and display geographical information in different ways. In many cases it is not sufficient to display the data but to analyse them so that it is possible to:

i   compare different sets of data which have been collected;
ii compare the results with other published findings;
iii test whether certain items of data are related or not related to each other.

Unless the data are expressed statistically, it may not be possible to say, for example, that location A receives more rainfall than location B. By calculating the *arithmetic average* it is possible to see whether a difference exists and by how much one station receives more than the other. If some additional calculations are made, it is possible to see if the difference between the two stations is *statistically significant*.

This section of the book is designed to show you both how to calculate the statistics and to demonstrate the types of situation in which particular statistical tests should be used.

It is most important when carrying out statistical analyses of any form to pay particular attention to:

i   the accuracy of the calculations;
ii the choice of statistical test.

Clumsy errors in calculation can be avoided by using one of many microcomputer packages, but care must be taken to input the data carefully. You will still need to know which test to perform in the first instance and how to interpret the result. We assume that you do not have access to computers and all calculations are given here in longhand. One important reason for giving the formula and working out statistics by hand is that you will get a much better understanding of what the calculations mean. Even if you have access to a computer, it is recommended that you try all of these tests by hand on a simple set of data, in order to familiarise yourself with the approaches.

5.2 STATISTICAL ANALYSIS

Statistical analyses are broadly divisible into two major types called *descriptive* and *inferential statistics*, both of which can be represented by *statistical notation*.

i   *Descriptive statistics* are calculated in order to summarise the properties of a set of data. For example, the arithmetic average or mean of a data set is a useful summarising measure. This is not the only measure of *central tendency* as we will see in a later section. Similarly, if we want to know something about the spread of values in a set of data, the *quartile ranges* or *standard deviation* might be calculated.

ii *Inferential statistics* are designed to test hypotheses or ideas we might have about the way in which one variable is related to another. For example, we can test to see if there is a relationship between soil depth and the angle of slope, or how a trend in one variable, such as unemployment, might be related to a trend in a second variable, such as the incidence of reported crime. We will examine the problems of setting up hypotheses, carrying out statistical tests and drawing conclusions from these tests at a later stage.

iii *Statistical notation*. Statisticians have developed a language, or mathematical notation, in order to simplify the instructions given for performing calculations. This basic notation is unavoidable and you should make every effort to become familiar with the relatively small number of symbols which are used. With reference to the following data, the two statements in Box 1 mean the same thing.

BOX 1   Mathematical symbols

Data   12 23 32 15 7 9

1 Add up all of the six numbers and divide the result by six to calculate the arithmetic mean.

$$\textbf{2} \quad \bar{x} = \frac{\sum_{i=1}^{n} x_i}{n}$$

Within the second statement in Box 1 are a number of symbols which will be used consistently throughout this chapter:

n represents the number of observations in the sample, in this case 6.

$\bar{x}$ means the *arithmetic mean* (or just *mean*)

x_i is each observation (i) in the data set (x)

$\sum_{i=1}^{n}$ *(sigma)* add up or sum the numbers from i = 1 (the first number in the list) to n (the last number in the list, in this case number 6).

Where more complex calculations are needed, the advantages of using the formula becomes even more obvious as we will see later.

The following two sections illustrate a range of descriptive statistics which can be applied to both numerical and spatial distributions. Any new notation or symbols will be explained at the appropriate point. You will also find a list of symbols in Table 5.1

TABLE 5.1
Symbols and abbreviations in statistical notation

Symbol			Symbol		
n	=	sample size	**$\bar{x}nn$**	=	mean nearest neighbour
x_i	=	each individual observation (i) in data set x	**NN_{ran}**	=	nearest neighbour statistic for a random distribution
$\sum_{i=1}^{n}$	=	sum all values from the first (i = 1) to the last (i = n)	**NN_{ind}**	=	nearest neighbour index
$\bar{x}$	=	arithmetic mean	**G**	=	Gini coefficient
$\hat{x}$	=	median	**LQ**	=	location quotient
$\ddot{x}$	=	mode	**χ^2**	=	chi-squared statistic
			E	=	expected frequency
Q_r	=	quartile range	**O**	=	observed frequency
Q_1	=	25th percentile (lowest quartile)	**df**	=	degrees of freedom
Q_2	=	50th percentile	**r_s**	=	Spearman's rank correlation coefficient
Q_3	=	75th percentile (highest quartile)	**r**	=	Pearson's product-moment correlation coefficient
s^2	=	variance	**t**	=	Student's t statistic
s	=	standard deviation			
CV	=	coefficient of variation			
CV%	=	percentage coefficient of variation			
SD	=	standard distance			

5.3
DESCRIPTIVE STATISTICS

Descriptive statistics are calculated in different ways depending on whether numerical or spatial distributions are being analysed. The difference is simply that in a numerical distribution we are dealing with

TABLE 5.2
Levels of measurement related to descriptive statistics of central tendency and dispersion

Level of measurement	Central tendency	Dispersion
Nominal	Mode	% in the mode Range
Ordinal	Median	Percentiles
Interval and Ratio	Mean	Standard deviation Variance

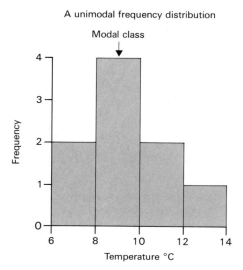

A unimodal frequency distribution

Modal class

Figure 5.1
Histogram drawn from the temperature data of table 5.3

TABLE 5.3
Midland temperature data arranged in groups

Groups (°C)	Values (°C)	Frequency
6–7·9	6·0, 7·4	2
8–9·9	9·5, 8·6, 9·6, 8·8	4
10–11·9	10·3, 10·4	2
12–13·9	13·7	1

tables of data. Descriptive statistics for spatial distributions are calculated for points located on a map or plan.

Numerical distributions

Many of the calculations we make are conditioned by the level of measurement used to collect the data in the first instance (Chapter 1). For example, the arithmetic mean of a set of data can only be calculated for data measured on the interval and ratio scales, and other measures of centrality and dispersion are needed for data measured on the nominal or ordinal scales (Table 5.2).

Continuous and discrete distributions. Not only are data distinguished on the basis of levels of measurement but we can also conveniently distinguish between *continuous* and *discrete* data. Continuous data are ungrouped data measured on the interval or ratio scale. Discrete data are grouped into mutually exclusive groups. For example, if several lengths are measured, we may estimate the number of times (frequency) certain distances fall within each group.

The nine data points below are daily temperatures (°C) for Midland England taken on successive days in April.

7·4 10·3 9·5 8·6 13·7 9·6 6·0 10·4 8·8

These data are measured on the interval scale and are continuous (i.e. they lie on a continuously changing scale, in this case at increments of 0.1°C). They can be turned into discrete data sets by selecting groups of temperatures. In Table 5.3 the data have been arranged into 4 groups. By arranging the values in these groups, the data are now said to be discrete, since from the frequency information we only know the number of times values between two limits occur and the exact value is unknown. Of course, we would not normally see the column of values but only the frequencies.

The histogram. One of the most common ways of showing discrete data grouped into frequency classes is to plot a *histogram*. The y or vertical axis of this graph represents the frequency or number of times values fall in each group. The x or horizontal axis represents the groups. In Figure 5.1 groups are based on a temperature scale. For each group, a bar graph with a length proportional to the frequency of observations in each class is drawn. In this way the histogram provides a visual display of the form of the data set and this can be very useful when choosing appropriate techniques for further statistical analysis.

Measures of centrality – mean, median and mode. Often it is useful to provide a single value which summarises a data set. There are three commonly used methods: mean, median, mode. The measure of centrality we calculate depends on the level of measurement (Table 5.2). For example the *arithmetic mean* of the above temperature data is calculated from the formula given in Box 1, i.e.

$$\bar{x} = \frac{6·0 + 7·4 + 9·5 + 8·6 + 9·6 + 8·8 + 10·3 + 10·4 + 13·7}{9} = 9·37$$

The *median* ($\hat{x}$) is another measure of centrality and represents the middle number in a set of ranked observations. In order to calculate the median it is important to arrange the data in *rank order* from the highest to the lowest value. In the above data set, the median is 9.5, since four temperatures are higher than and four temperatures are lower than this value. With an even number of observations, the average of the middle pair of observations is taken, e.g. in the following set

4 6 8 10

the median is $\frac{6 + 8}{2} = 7$

(Note that the median does not necessarily have to be a member of the data set.)

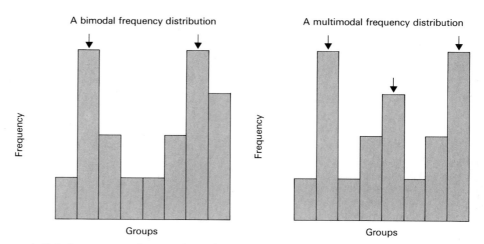

A bimodal frequency distribution

A multimodal frequency distribution

Figure 5.2
Bimodal and multimodal distributions

A third measure of central tendency often used for grouped data is the *mode* ($\bar{x}$) or modal class. For the data arranged into class intervals as shown in Table 5.3, the modal class is the group with most observations, in this case class 8–9.9°C which contains four values. The modal class for these data is also marked on Figure 5.1.

In some cases, there may not be just a single peak in a frequency distribution to give a *unimodal* distribution, but two or more *(bimodal or multimodal)* distributions may exist. Unimodal and bimodal properties are useful in some areas of physical geography, especially in studies of particle size distributions where, more often than not, the data may be bimodal or multimodal. Examples of a bimodal and multimodal frequency distribution are given in Figure 5.2.

Measures of dispersion – range, percentiles, variance and standard deviation. The mean, median and mode, when used on their own, provide no information about the range or spread of data. A number of measures of dispersion inform us about this. In the same way that some measures of centrality can only be calculated for nominal and ordinal scale data, the same condition also applies to measures of dispersion. These restrictions are summarised in Table 5.2.

The range. Of the several measures available, the *range* is by far the simplest and can be calculated for all levels of measurement. It is defined as the difference between the largest and smallest number in the data set, or

$$\text{range} = x_{max} - x_{min}$$

For the rainfall data given in Table 5.4, the range is 72.0 mm (December 72.9 minus July 0.9 mm).

The range is useful because it gives a quick guide to the dispersion of the data but of course it is very sensitive to the presence of any extreme values in a set of data (Table 5.5).

Percentiles. Of more use in cases where extreme values are to be found is the *percentile deviation* which can be calculated from ordinal data (or interval and ratio data if suitably rearranged). One of the most common percentile measures is the inter-quartile range, which is the range covering the middle half (50%) of a set of ranked data. In a list of ranked values, 25% of the numbers fall below the value of the 25th percentile or lower quartile (Q1), and 25% of the values fall above the value of the 75th percentile or upper quartile (Q3) (Table 5.5). Although many different percentile ranges may be calculated (e.g. quintiles (5), sextiles (6)), only the inter-quartile range will be considered here.

If the number of values in a data set, such as the 7 numbers listed below, is not divisible by 4 to provide an integer value (whole number), the exact value of the 25th and 75th percentiles must be calculated:

23 24 34 43 45 56 89
     ↑   ↑
    Q1   Q3

TABLE 5.4
Monthly rainfall data for North Africa (mm)

Station at Sidi Kacem, Morocco	1915–1983 average.
Jan	65.9
Feb	62.6
Mar	62.2
Apr	49.9
May	27.2
Jun	12.1
Jul	0.9
Aug	1.6
Sep	11.1
Oct	43.4
Nov	70.9
Dec	72.9

TABLE 5.5
Definition of simple measures of dispersion

	Ranked Value	Rank
	73.7	1
	62.1	2
	58.7	3
Upper	38.4	18
Quartile	37.9	19
(Q3)	32.1	20
Median	20.4	35
(Q2)	19.8	36
	17.4	37
Lower	13.6	52
Quartile	12.8	53
(Q1)	12.4	54
	10.1	70
	9.9	71
	9.8	72

inter-quartile range

The lower quartile (Q1) lies somewhere between 24 and 34 and the upper quartile (Q3) somewhere between 45 and 56. The exact number is found by using a weighted value calculated from the formula given in Box 2. (Note the only value which changes in this calculation is the sample size for a given data set.)

BOX 2   Calculation of upper and lower quartiles

The weighting for the lower quartile value (Q1 or 25th percentile) is

$Q1 = n/4 + 0.5 = 7/4 + .5 = 2.25$

For the upper quartile value (Q3 or 75th percentile) the weighting is

$Q3 = (3 \times n)/4 + .5 = 21/4 + .5 = 5.75$

where n = is sample size.

The weighting values calculated in Box 2 need careful interpretation. The first digit provides a position from the lowest value. In this case **2·(25)** indicates that Q1 will be greater than the second value in the data set (24) but less than the third (34). The number after the decimal point (2)·**25** tells us how far Q1 lies after the second value:

$Q1 = 24 + 0.25 \times (34 - 24) = 26.5$

$Q3 = 45 + 0.75 \times (56 - 45) = 53.25$

In this case, the inter-quartile range (Q_r) is

$Q_r = Q3 - Q1 = 53.25 - 26.5 = 26.75$

If the number of observations in the data set is exactly divisible by 4, the calculation of the upper and lower quartiles is unnecessary. For example, with the 12 numbers listed below, 3 values lie above and below the quartile range:

12 14 16 18 22 24 25 33 35 39 41 51
            ↑   ↑
            Q1   Q2

The inter-quartile range is:

$35 - 18 = 17$

High inter-quartile ranges relative to the mean or median of a data set tells us that there is a large spread of values, whereas low values relative to the same measures of centrality tells us that the data are clustered.

Similar calculations can be made for other percentile ranges although this is not common in geographical studies in general. An exception in geomorphology is found when analysing the size distribution of sediments, when a variety of percentiles are often calculated.

Variance and standard deviation. These measures of dispersion take account of *all* values in a data set and are some of the most useful summarising statistics. They both measure the spread of values around the mean, but can only be used:

i for data measured on interval or ratio scales;
ii for data sampled randomly from a population with an approximately *normal* frequency distribution (see page 112).

The variance (s^2) is simply the square of the standard deviation. The variance is expressed as:

$$s^2 = \frac{\sum_{i=1}^{n} (xi - \bar{x})^2}{n - 1}$$

The standard deviation (s) is the square root of the variance:

$$s = \sqrt{s^2}$$

To calculate the variance, we have to:

1 take the difference between each individual data point and the mean;
2 square the result and add up all of the squared values;
3 divide the resultant sum by the sample size minus 1.

The upper term in the formula used to calculate the variance and standard deviation is often easier for calculation on a calculator or computer by using a special computational formula. This formula is given in Box 3.

For the five numbers listed below, we can calculate the variance and standard deviation from this quicker method in the following way:

1 add up the values of x and x^2 as shown below;
2 square the sum of x;
3 substitute these values into the quick computing formula of Box 3 as shown below.

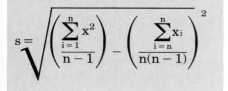
x	x^2
3	9
6	36
8	64
7	49
5	25

$$\Sigma x = 29 \quad \Sigma x^2 = 183$$

$$(\Sigma x)^2 = 841$$

$$n - 1 = 5 - 1 = 4$$
$$n(n - 1) = 5(5 - 1) = 20$$

Substituting these values into the quick formula of Box 3 gives:

$$s = \sqrt{183/4 - 841/20} = \sqrt{45{\cdot}75 - 42{\cdot}05} = \sqrt{3.7} = 1{\cdot}924.$$

For these five values, we have:

Variance (s^2) = 3·7
Standard deviation (s) = 1·924

As with the inter-quartile range, a high standard deviation and variance means that the data set has a large range of values, whereas the data are grouped closely around the arithmetic mean ($\bar{x}$) when the values are small.

Comparing the variation in different data sets – coefficient of variation. It is often useful to compare two or more groups of data, perhaps collected from different areas, to see not only which has a higher mean but which set of data has a greater spread of values. For example, look at the statistics for two rainfall stations given below:

Station	mean (mm)	standard deviation (mm)
1	642	75
2	437	68

At first glance, we might be tempted to say that station 1 has both a larger mean and a larger spread of values, but is this strictly true? The reason this may not be the case is because the absolute values at both stations are different and the standard deviation simply reflects the actual numbers in the data set (a problem which also occurs when we make similar comparisons of data measured on different scales or in different units). This problem is overcome by calculating the *coefficient of variation*. This is defined as:

$$CV = s/\bar{x}$$

where s = the standard deviation
$\bar{x}$ = the arithmetic mean.

It is sometimes expressed as a percentage (CV%) by multiplying the result by 100.0.

In the case above,

Station 1   CV = 75/642 = .1168   CV% = 11.68%
Station 2   CV = 68/437 = .1556   CV% = 15.56%

We can see that our initial conclusion is incorrect, and that although station 1 has a higher mean, the variability in the data from station 2 is greater.

Parametric and non-parametric statistics – the normal curve. When the frequency of observations in particular class intervals are plotted in the form of a histogram, they possess a property known as a frequency distribution. There are many types of frequency distribution (Figure 5.2). There is one type of frequency distribution which is of special importance and which can be precisely described in terms of its mean and standard deviation. In general form this distribution is shaped like a bell, and is symmetrical about its mean, median and modal class (Figure 5.3). Also shown is what happens to the histogram when we divide the class intervals into very small units. Here, instead of a histogram we have a line forming a continuous bell shaped curve known as the *normal curve*. The normal curve has several important properties (see Figure 5.4):

i 68% of all the values in the frequency distribution lie within the range +/− 1 standard deviation unit;
ii 95% of all the values in the frequency distribution lie within the range +/− 2 standard deviation units;
iii 99% of all the values in the frequency distribution lie within the range +/− 3 standard deviation units.

**Figure 5.3
A normal frequency distribution and normal curve**

For example, if we had 200 observations in a set of data and grouped them into frequency classes, 136 of them would lie between +/− 1 standard

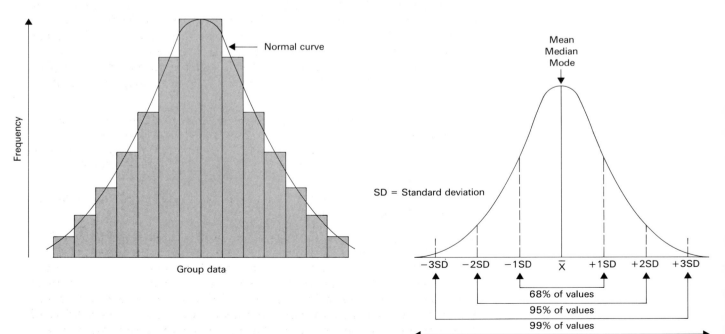

**Figure 5.4
Properties of the normal curve**

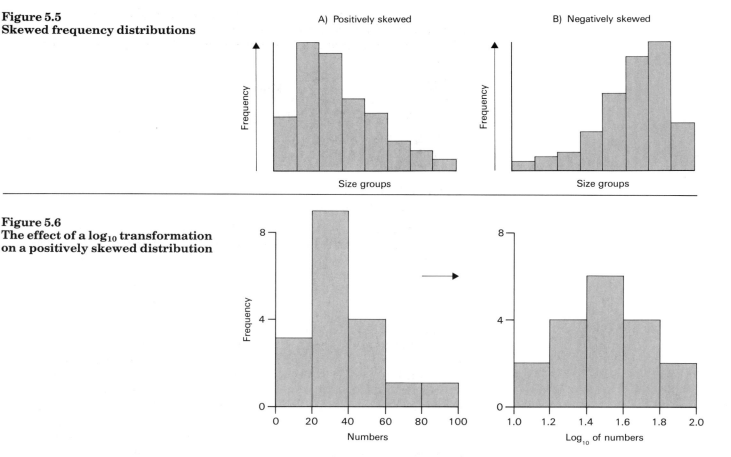

Figure 5.5
Skewed frequency distributions

A) Positively skewed

B) Negatively skewed

Figure 5.6
The effect of a $\log_{10}$ transformation on a positively skewed distribution

deviation from the mean and so on. Of course, many distributions do not look like the normal curve. Some, like the two given in Figure 5.5, are not symmetrical. These are called *skewed* distributions, and we can see that in such circumstances the mean, median and mode may occur in different places on the curve (remember that they are in the same place for a normal distribution). The normal distribution is one of the most important statistical distributions because many of the inferential statistics that will be described in later sections assume that the data have this frequency distribution. All statistics which assume this distribution are known as *parametric statistics*. These make up the most powerful tests of association and difference, as we will see later. However, if you plot the frequency distribution of a data set and find that it does not have a normal frequency distribution, then there are two alternatives available. First, you could change (*transform*) your data set so that it becomes normal. For example, look at the histograms of Figure 5.6. This change is accomplished in the following way:

i   for each value in the data set, find the logarithm of the number (log base 10). Most calculators can perform this simple function;
ii reclassify the new data into the same number of groups as shown on the x axis, but with a different range to cover the logarithm of each number;
iii count the number of values which fall in each of the new class intervals and plot these as a histogram.

The positively skewed distribution has become *normal* as in Figure 5.6. This transformation of data does not always give rise to a normally distributed set of values. If in doubt, or if you have small sets of data and are unable to find out whether or not your data are normally distributed, it is advisable to use a second type of test called a *non-parametric test*. Non-parametric tests make no assumptions about how the data are distributed. Again, there are many tests available, and in a later section you will be shown some options available to you for different applications and under different conditions. These are summarised in Table 5.19.

113

EXERCISE 1:   MEASURES OF CENTRALITY AND DISPERSION

TABLE 5.6 Annual rainfall for a station at Sidi Kacem, Morocco, from 1915 to 1980

Year	Rainfall mm	Year	Rainfall mm	Year	Rainfall mm	Year	Rainfall mm	Year	Rainfall mm	Year	Rainfall mm
1915	372·6	1927	577·7	1939	481·7	1951	571·1	1963	635·9	1975	345·0
1916	380·0	1928	475·5	1940	516·0	1952	562·1	1964	452·0	1976	603·4
1917	638·5	1929	307·3	1941	597·0	1953	429·5	1965	614·3	1977	469·0
1918	485·2	1930	492·1	1942	638·9	1954	353·6	1966	359·2	1978	499·6
1919	455·0	1931	363·4	1943	414·1	1955	636·1	1967	430·5	1979	672·2
1920	397·3	1932	385·0	1944	389·4	1956	548·9	1968	585·3	1980	730·6
1921	378·7	1933	621·9	1945	253·6	1957	582·3	1969	704·0		
1922	536·6	1934	411·6	1946	440·6	1958	459·4	1970	375·6		
1923	499·1	1935	313·3	1947	490·4	1959	380·3	1971	674·6		
1924	444·3	1936	743·1	1948	343·0	1960	650·4	1972	422·4		
1925	370·8	1937	454·7	1949	497·6	1961	349·2	1973	377·3		
1926	570·9	1938	435·8	1950	362·5	1962	565·9	1974	340·4		

Q **1** Calculate the mean, variance and standard deviation of this data set.

2 Calculate the mean, variance and standard deviation of the rainfall data for England and Wales given in Table 3.5 for the period 1915–1980.

3 Calculate the coefficient of variation for both data sets and decide which location has the highest mean annual rainfall and the least reliable rainfall.

4 Group the data from both data sets into the following frequency classes:

TABLE 5.7
Class intervals for rainfall analysis

Rainfall range mm	No. of years Morocco	No. of years England and Wales
100– 199		
200– 299		
300– 399		
400– 499		
500– 599		
600– 699		
700– 799		
800– 899		
900– 999		
1000–1099		
1100–1199		

From the data of Table 5.7:

5 Plot a histogram of the frequency distributions and shade the locations in different patterns for comparative purposes, as shown in Figure 5.7.

6 Explain whether you think one or both of these data sets have a normal frequency distribution.

7 Identify the modal class and mark the position of the modal class for each data set on the histogram.

8 Find the median value for each data set.

9 Calculate the inter-quartile ranges for each data set.

10 Tabulate the statistics (see Table 5.8).

TABLE 5.8	Measures of centrality	Morocco	England and Wales
	Mean		
	Median		
	Modal class		
	Measures of dispersion		
	Range		
	Inter-quartile range		
	Variance		
	Standard deviation		

From Table 5.8 and the histogram:

11   What do these descriptive statistics reveal about the rainfall characteristics of these two locations?

12   How useful are these descriptive statistics for describing rainfall characteristics?

Figure 5.7
A sample histogram for two stations showing modal classes

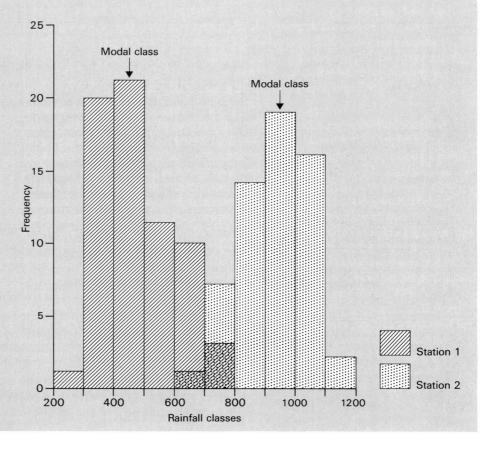

Spatial distributions
In the last part of this chapter we considered measures of centrality and dispersion which related to numerical distributions. This section considers the distribution of data in space and takes a similar approach to that adopted in the previous section.

When data are set out in spatial rather than tabular form, such as the distribution shown in Figure 5.8, it seems reasonable that we would be able to define the centre of such a distribution. In order to define the value quantitatively, however, it is necessary to place a grid over the spatial distribution, and decide on:

 i the quantitative scale of the grid to be used;
ii the orientation of the grid.

115

Figure 5.8
Analysis of spatial distribution to locate mean centre ($\bar{x}_c$) and weighted mean centre ($\bar{x}_w$)

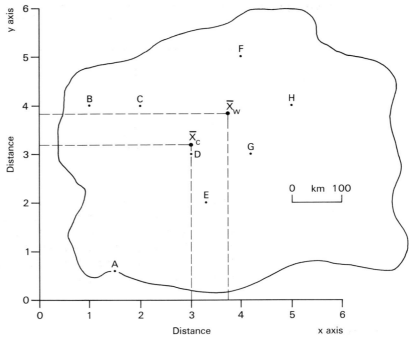

Figure 5.9
Definition of X and Y co-ordinates

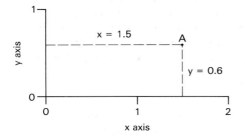

TABLE 5.9

Point	x axis	y axis
A	1·5	0·6
B	1·0	4·0
C	2·0	4·0
D	3·0	3·0
E	3·3	2·0
F	4·0	5·0
G	4·2	3·0
H	5·0	4·0
Totals	24.0	25.6

The choice of scale is important in that it defines the magnitude of the distances along each axis. If two or more distributions are to be compared, for example, it is best to ensure that the same scales are used for all cases. The measures of centrality and dispersion also depend on the orientation of the axes. On maps, it is convenient to use the Ordnance Survey grid as the basis for orientation. On aerial photographs, or other sources without a superimposed grid, geographical north could be used to orient the grid. It is usual to superimpose the grid in such a way that its origin lies to the south-west of a distribution (Figure 5.8). Once the grid has been superimposed and scaled, the quantitative measures may be obtained.

Measures of centrality – mean, median and modal centres.

Mean centre. All the values plotted on Figure 5.8 can be represented by coordinate distances along the x and y axes. The values in Table 5.9 are derived as for the first pair of coordinates in Figure 5.9.

The next step is to calculate the mean of both sets of data as we did for the numerical distributions (Box 1).

The mean of the x coordinates is $\bar{x} = 24\cdot0/8 = 3\cdot0$

The mean of the y coordinates is $\bar{y} = 25.6/8 = 3.2$

These two values, when plotted on the original distribution, give us the mean centre ($\bar{x}_c$) of x and y as shown on Figure 5.8.

The calculation shown above, however, assumes that all points in the spatial pattern are equally important. However, if we assume that the same points on Figure 5.8 represent towns with different populations we may want to find the mean centre of the distribution weighted by the population of each settlement (Table 5.10).

For the same coordinates as above, but now using a figure for population to weight the importance of each settlement, the calculation is as below, using the figures from Table 5.10.

The mean weighted centre is found from the following calculations:

$$x_w = \sum_{i=1}^{n} xw / \sum_{i=1}^{n} w = 962/256 = 3\cdot75$$

$$y_w = \sum_{i=1}^{n} yw / \sum_{i=1}^{n} w = 981/256 = 3\cdot83$$

The result is that the weighted mean centre moves much closer to towns F and H which have considerably higher populations (Figure 5.8).

116

TABLE 5.10

Point	x coordinate	y coordinate	population (w) thousands	x * pop (xw)	y * pop (yw)
A	1·5	0·6	10	15.0	6·0
B	1·0	4·0	25	25·0	100·0
C	2·0	4·0	12	24·0	48·0
D	3·0	3·0	9	27·0	27·0
E	3·3	2·0	30	99·0	60·0
F	4·0	5·0	70	280·0	350·0
G	4·2	3·0	10	42·0	30·0
H	5·0	4·0	90	450·0	360·0
Totals			256	962.0	981.0

Median centre. Just as the median can be calculated for numerical distributions, it may also be calculated for point pattern distributions. The median centre is a point of intersection where 50% of the data set lie above and 50% below the value, on each of the x and y axes (point $\bar{x}_{c1}$, Figure 5.10). Its advantage over the mean centre is that its location can be found very quickly without calculation.

Figure 5.10
The effect of different grid orientations on the location of the median centre

$\overline{X}_{C1}$ = median centre from orthogonal axes

$\overline{X}_{C2}$ = median centre from 45° axes

As suggested in the introduction, the exact position of this point depends on the orientation of the grid. If we superimpose a grid at 45° to the original x and y axes in Figure 5.8, the position of the median centre will also move (point $\bar{x}_{c2}$, Figure 5.10).

Modal centre. The modal centre of a point pattern corresponds to the mode of the numerical distribution. In a spatial sense, it is the area in which the largest number of points lie. In the point pattern distribution of Figure 5.11 an arbitrary series of grid squares has been superimposed on the map. The number of observations falling in each square is then identified, and the modal centre is the square for which most observations occur. Like the other measures of centrality the position of the modal centre is sensitive to the scale and orientation of the grid. Too small a grid might give rise to a multi-modal distribution whereas too large a grid produces a large area representing the modal centre.

The choice of mean, median and modal centres is conditioned by the same criteria, i.e. levels of measurement and normality, that apply to the mean median and mode of numerical distributions (Table 5.2).

A measure of dispersion – standard distance. The standard distance is a measure which tries to describe the spread of data in a point pattern and is

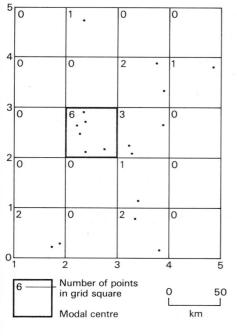

6 ———— Number of points in grid square

0 ———— 50

Modal centre   km

Figure 5.11
Definition of the modal centre

similar to the standard deviation in a numerical distribution. The calculation of standard distance is represented in terms of the formula given below.

$$SD = \sqrt{\frac{\sum\limits_{i=1}^{n} d^2}{n}}$$

where:

    d = the distance of each point from the mean centre (measured in the unit scale of the grid)

    n = the number of points

    A quick computing formula for the standard distance can also be used as shown in Box 4.

BOX 4   Quick computing formula for the standard distance

$$SD = \sqrt{\left(\frac{\sum\limits_{i=1}^{n} x^2}{n} - \bar{x}^2\right) + \left(\frac{\sum\limits_{i=1}^{n} y^2}{n} - \bar{y}^2\right)}$$

where: n = sample size

              $\bar{x}$ = mean of x

              $\bar{y}$ = mean of y

For the table of coordinates used to calculate the mean centre (Table 5.9), the standard distance is calculated as follows:

1 sum the values of the x and y coordinates;
2 calculate the mean centre for x and y;
3 sum the squared values of the x and y coordinates (Table 5.11);
4 substitute the resulting values into the quick computing formula of Box 4.

TABLE 5.11

Station	x	x^2	y	y^2
A	1.5	2.25	0.6	0.36
B	1.0	1.00	4.0	16.00
C	2.0	4.00	4.0	16.00
D	3.0	9.00	3.0	9.00
E	3.3	10.89	2.0	4.00
F	4.0	16.00	5.0	25.00
G	4.2	17.64	3.0	9.00
H	5.0	25.00	4.0	16.00
	Σ24.0	Σ85.78	Σ25.6	Σ95.36

$$\bar{x} = \frac{24 \cdot 0}{8} = 3 \cdot 0 \qquad\qquad \bar{y} = \frac{25.6}{8} = 3.2$$

$$\bar{x}^2 = 9 \cdot 0 \qquad\qquad\qquad \bar{y}^2 = 10.24$$

By substituting the above values into the quick computing formula given in Box 4:

$$SD = \sqrt{\left(\frac{85 \cdot 78}{n} - 9 \cdot 0\right) + \left(\frac{95 \cdot 36}{n} - 10 \cdot 24\right)}$$

$$SD = \sqrt{1 \cdot 72 + 1 \cdot 68}$$

$$SD = \sqrt{3 \cdot 4}$$

$$SD = \quad 1 \cdot 84$$

118

Figure 5.12
Plotting the standard distance

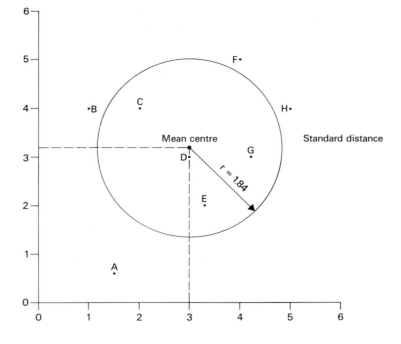

The value of the standard distance, 1.84 (in the grid units used), is the radius of a circle which should be drawn about the mean centre to show the area in which 68% of all values would fall. The standard distance for the distribution shown in Figure 5.8 is plotted on Figure 5.12.

The size of circle plotted on the map reflects the spread of the data. A large circle suggests that the points are widely distributed about the mean centre whereas a small circle suggests that the data are more clustered around the mean centre. Remember that if you are comparing two different distributions, the same units must be used because the radius of the circle is a function of the grid scale.

Measures of spatial patterns

Measures of central tendency and dispersion enable us to describe objectively some elements of the patterns within spatial distributions. However, when we talk of *settlement patterns* or *land use patterns* the word *pattern* suggests that we know something about the regularity or irregularity of the distribution of points. Of course, measures like the mean centre and standard distance do not allow us to make judgements about these aspects of spatial patterns. Other statistical techniques are available which enable us to comment in this way.

A measure of regularity – nearest neighbour analysis. The nearest neighbour index gives a quantitative measure of the arrangement of points in space. On the basis of a simple calculation it is possible to say with some degree of certainty whether a distribution is clustered, randomly or regularly arranged in space.

There are a number of steps in the calculation of the nearest neighbour index. For the point pattern of Figure 5.13:

1 Identify the nearest neighbour of each point. These are given in Table 5.12 along with the straight-line distances between them;

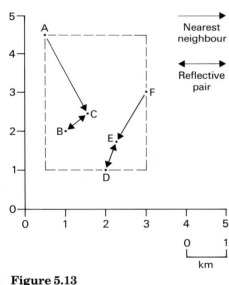

Figure 5.13
Nearest neighbour analysis

	TABLE 5.12	Point	Nearest Neighbour	Distance (km)
		A	C	2.3
		B	C	0.7
		C	B	0.7
		D	E	0.8
		E	D	0.8
		F	E	1.5
			Total	6.8

119

2 Add up all of the distances and divide the result by the number of points in the distribution to calculate the mean nearest neighbour distance (Box 5);

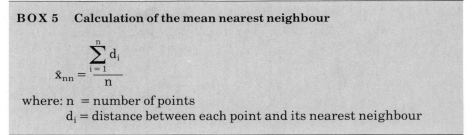

BOX 5   Calculation of the mean nearest neighbour

$$\bar{x}_{nn} = \frac{\sum_{i=1}^{n} d_i}{n}$$

where: n = number of points
d$_i$ = distance between each point and its nearest neighbour

From the above formula, the mean nearest neighbour distance for the data given in Figure 5.13 is

$$\bar{x}_{nn} = 6 \cdot 8/6$$

$$= 1 \cdot 13$$

3 In order to be able to establish whether the pattern is what we would expect to get from the same number of points distributed randomly, we need to know the area from which the data are derived. In Figure 5.13, the area is $2 \cdot 5 \times 3 \cdot 5$ km, or 8.75 km^2. This value is used to calculate an expected value for a random distribution according to the formula of Box 6;

BOX 6   Calculation of a random nearest neighbour

$$NN_{ran} = 1/\left(2 \times \sqrt{p}\right)$$

where: p = the density of points (number per unit area)

In the above example, $p = 6/8 \cdot 75 = \cdot 686$

$$NN_{ran} = 1/\left(2 \times \sqrt{\cdot 686}\right)$$

$$= \cdot 604$$

4 On the basis of the calculated and random nearest neighbour, the next step is to calculate the *nearest neighbour index*. This index provides a single measure of the distribution on a scale ranging from 0 (clustered), through 1 (random) to $2 \cdot 15$ (dispersed). It is calculated by dividing the observed value by the value obtained for a random distribution. In this case

$$NN_{ind} = \frac{\bar{x}_{nn}}{NN_{ran}}$$

$$NN_{ind} = 1 \cdot 13/ \cdot 604$$

$$= 1 \cdot 871$$

On the above scale, we can see that the distribution of points in Figure 5.13 is closer to a dispersed than a random distribution.

One of the major limitations of this statistic is the arbitrary way in which the grid is placed upon the spatial distribution. Look at the distribution of Figure 5.14, for example. If the whole area (200×100m) is used to calculate the nearest neighbour index, the value is $1 \cdot 1$, close to a random distribution. If the total area is subdivided into regions A and B on Figure 5.14, however, the indices become $1 \cdot 9$, approaching dispersed, and $0 \cdot 4$, approaching clustered, respectively. Try to ensure that the way in which you sample the data does not lead to similar problems in interpretation. A quick check on this is to calculate the nearest neighbour index for different size grids to ensure that all values are similar.

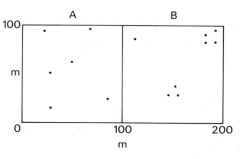

Figure 5.14

Measures of concentration – the Gini coefficient and Lorenz curve. Another approach to the problem of describing spatial patterns is provided by an index of concentration kown as the *Gini coefficient* and the graphical display of the results from such analysis through a diagram known as the *Lorenz curve.* This statistic may be used to analyse numerical distributions as well as to describe associations between point patterns. There are many applications. For example, we could see whether regions of the UK had lower or higher employment in different industries than the UK as a whole, or we could look at the distribution of wealth in the same way. The statistic could be used to look at the equality of wealth distribution per capita on a national or internatonal scale and/or by ethnic group or social class. A further application is that the Gini coefficient can be calculated for the same data set for different years to see whether differences are becoming greater or smaller through time.

The Gini coefficient is calculated from the formula given in Box 7.

BOX 7   Calculating the Gini coefficient from two data sets

$$G = 0.5 \times \left(\sum_{i=1}^{n} |X_1 - Y_1| \right)$$

The values of X_1 and Y_1 are the percentage points of two numerical or spatial distributions.

The vertical brackets around these values are not the same as in previous calculations, and mean the *modulus of,* i.e. the difference between the two percentage distributions irrespective of sign.

In the following example, an attempt is made to establish whether similar proportions of employees are to be found in all geographical regions. In other words, does the percentage of employees in the retail sector reflect the regional distribution of employees in services?

1 take the difference, irrespective of sign ($+/-$) between the percentage of employees in the retail sector and that percentage employed in total service employment;
2 sum the differences;
3 multiply the result by 0·5.

By substituting these values into the formula in Box 7, the Gini coefficient is:

$$G = 0.5 \times 70.6 = 35.3$$

The value of G will be on a scale of 0 to 100, with 0 indicating perfect association between the two frequency distributions and 100 indicating that they are as different as possible. The calculations above show that the two distributions of retail industry and service sector employment are only weakly associated. Note that this statistic is not affected by the normality or non-normality of the data. In addition to the presentation of a statistical measure alone, it is also possible to show the data graphically by rearranging it as a cumulative frequency in order to plot the *Lorenz curve.*

From the data of Table 5.13, the regions are ranked on the basis of the ratio between the percentage distribution of variables X and Y (X/Y) as shown in Table 5.14.

The pairs of cumulative X and Y values are then plotted on a graph as shown in Figure 5.15. The straight line between the two corners of the graph would be a perfect match between the two distributions (Gini coefficient of 0). The greater the departure from the line in this case means the greater degree of localisation or regional specialisation.

Location quotient. An alternative statistic which also shows the deviation of individual regions from a mean or national average is the *location quotient.* This can be used to show the concentration of a particular activity

TABLE 5.13

Percentage of employees in the retail sector (X) compared with the percentage of those in total service employment (Y)

Region	X (%)	Y (%)	X–Y (%)
London & SE	14·9	3·6	11·3
East and South	8·5	8·5	0·0
South West	9·6	5·2	4·4
W Midlands	6·6	15·5	8·9
E Midlands	19·5	10·1	9·4
Yorks & Humber	7·4	6·9	0·5
North West	23·7	14·0	9·7
North	0·9	4·5	3·6
Scotland	3·4	11·4	8·0
Wales	5·5	20·3	14·8
Total	100.0	100.0	70.6

TABLE 5.14

Region	X/Y	Cumulative X	Cumulative Y
London & SE	4·13	14·9	3·6
E Midlands	1·93	34·4	13·7
South West	1·85	44·0	18·9
North West	1·69	67·7	32·9
Yorks & Humber	1·07	75·1	39·8
East and South	1·00	83·6	48·3
W Midlands	0·43	90·2	63·8
Scotland	0·30	93·6	75·2
Wales	0·27	99·1	95·5
North	0·20	100·0	100·0

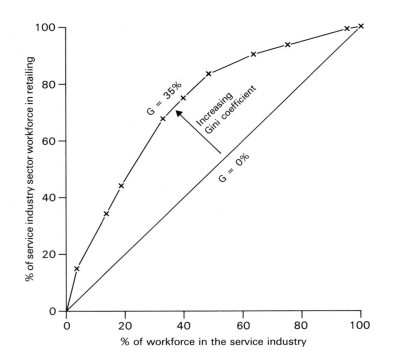

Figure 5.15
The Lorenz curve

or characteristic in an area. The values of X/Y listed in Table 5.14 are location quotients, showing:

$$LQ = \frac{\% \text{ of workforce in retailing}}{\% \text{ of workforce in service industry}}$$

Values greater than 1 indicate *more than expected*, whereas values of less than 1 indicate *fewer than expected*. This statistic can be applied in many situations. For example, the distribution of ethnic groups in different parts of a city may be analysed by comparing the total population with the proportion in different ethnic groups in different enumeration districts. In this way, patterns of ethnic segregation within a city can be highlighted:

$$LQ = \frac{\dfrac{\text{Population of ethnic group in ED}_i}{\text{Total population in ED}_i}}{\dfrac{\text{Total population of ethnic group in city}}{\text{Total population in city}}}$$

Similar applications can be made to look at the distribution of agricultural production, land use, wealth, and other spatially distributed phenomena. These data can be arranged in suitable groups and plotted as a choropleth map (Chapter 4).

EXERCISES: SPATIAL DISTRIBUTIONS

EXERCISE 1: **MEASURES OF CENTRALITY AND DISPERSION**

Figure 5.16 shows the location of settlements in the parish of Upham in Wilfordshire. At the parish council meeting it was decided to build a new community centre, but all councillors wanted the facility close to their own villages. You have been called in to act as umpire in this dispute and provide a report justifying where you would site the community centre. Use the data tabulated below (Table 5.15) which gives the coordinate points of the centre of each village on the map.

Q **1** Plot the position of the median centre on the map.

2 Calculate and plot the position of the mean centre and standard distance circle on the map.

3 Calculate and plot the position of the population weighted mean centre on the map.

4 Give your recommendations as to which of these points should be chosen for the new community centre.

5 Explain what other information you would like to have considered before making such an important decision.

TABLE 5.15
Parish of Upham: settlement locations

Village	X Coordinate	Y Coordinate	Population
A	1·5	3·5	240
B	2·5	4·5	1100
C	3·0	1·0	98
D	3·5	3·0	860
E	5·5	3·4	952
F	6·5	2·0	3500
G	8·0	2·0	750
H	9·0	4·0	28

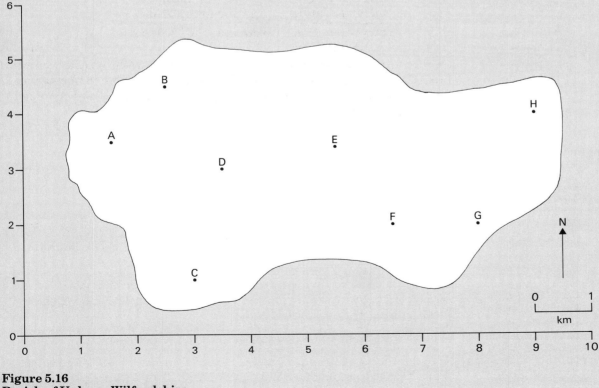

Figure 5.16
Parish of Upham, Wilfordshire

123

EXERCISE 2: MEASURES OF REGULARITY

The points plotted in Figure 5.17 represent the position of young trees in a survey of a small area of natural forest. You want to know whether these saplings are distributed regularly, randomly or in a clustered fashion in order to see whether germination has been equally successful over the whole area. The data in Table 5.16 gives the nearest neighbours of each tree in Figure 5.17.

Q **1** Calculate the mean nearest neighbour distance.

2 Calculate the expected nearest neighbour for a random distribution.

3 Calculate the nearest neighbour index.

4 Explain how your result may help you to understand the pattern of tree distribution.

5 Consider what other factors you might now need to explore to understand the problem better.

TABLE 5.16
Distribution of saplings

Tree Number	Nearest Neighbour	Distance (m)
1	2	0·500
2	1	0·500
3	4	0·425
4	3	0·425
5	4	0·550
6	5	2·000
7	8	1·075
8	9	0·450
9	8	0·450
10	11	0·625
11	12	0·400
12	11	0·400
	Total	7·8

Area of quadrat is 5m × 7m = 35 m².

Figure 5.17
Location of saplings in a 35m² quadrat

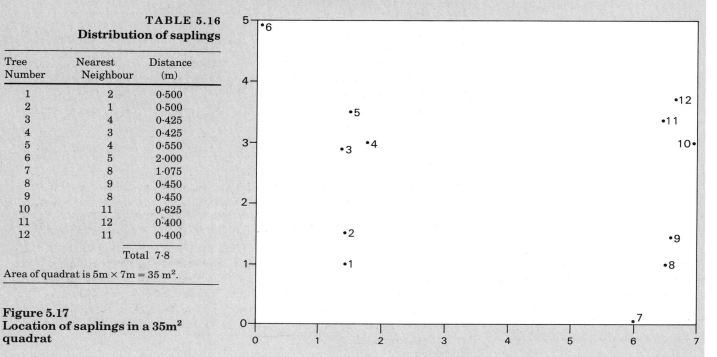

EXERCISE 3: MEASURES OF CONCENTRATION

TABLE 5.17
EC employment statistics

Country	% employed by country	% of EC's agricultural workforce
Belgium	2·9	1·0
Denmark	2·1	1·7
W Germany	20·7	13·5
Greece	3·0	10·0
Spain	8·6	17·0
France	17·3	15·3
Ireland	0·5	1·6
Italy	16·9	22·1
Luxembourg	0·5	0·1
Netherlands	4·2	2·4
Portugal	3·4	9·3
United Kingdom	19·9	6·0
Totals	100.0	100.0

SOURCE: Eurostat, 1985

The data in Table 5.17 are figures for the percentage of the total EC population in civilian employment by country, related to that country's percentage share of the EC's agricultural workforce.

Q **1** Calculate the Gini coefficient for these two distributions.

2 Calculate the location quotients for the data and plot the distributions in the form of a Lorenz curve.

3 Classify the location quotients into approximately 4 groups. Assign a method of shading to each group and produce a choropleth map to show variations in location quotients.

4 Explain the Lorenz curve and interpret the Gini coefficient.

5 With reference to the map of the EC countries, describe and account for the pattern of location quotients.

124

Introduction

In many geographical studies the data are collected with the aim of describing contrasts between regions. For example, we might collect data on crop yields for particular soil types or geological outcrops, temperature data from inside and outside urban areas, or pebble shape data in the upstream and downstream reaches of river systems. There are simple tests of significance which allow us to say in an objective way whether the differences between samples or areas could have occurred by chance. In order to do this, we must follow a number of stages:

i set up a hypothesis to test;
ii select a suitable test and collect data in accordance with the test assumptions;
iii decide on the confidence level we wish to impose on the test;
iv calculate the test statistic;
v accept or reject the hypothesis;
vi draw conclusions from the analysis.

These procedures will be explained in general terms with reference to a study of temperature contrasts between urban and rural areas.

Hypothesis testing. We want to know whether there is a difference between air temperature values in two adjacent areas, one urban and one rural. An inferential statistical test is to be used which makes no assumption about the frequency distribution of the data but assumes that the data are independent and are collected randomly from the population (Chapter 1).

Because we are trying to be objective, the hypothesis we actually test is called the *null hypothesis*. This states that:

'There is no significant difference between temperatures within and outside the urban area.'

This may not be true, and we always have another hypothesis available before carrying out the test. In this case, the alternative hypothesis states:

'There is a significant difference between temperatures within and outside the urban area.'

At the beginning of statistical tests of inference it is good practice to state the null and alternative hypotheses because, if the first is rejected, an alternative should always be available *before the test is made*. Note also that it is the null or negative hypothesis which is being tested, never the alternative hypothesis. In other words, we always 'bend over backwards' to try and prove ourselves wrong!

Following the setting of hypotheses, collect the data either from primary or secondary sources. For example, the data given in Table 5.18 shows the number of times on which midnight temperatures fall within a certain range of values.

The exact nature of the test will depend on the data available. This point will be considered later.

Confidence levels. Because the analysis is performed on a sample, rather than all occasions on which the data could have been collected, we can never be totally certain that the conclusion drawn is correct. If this is expressed on a percentage scale, it means that we are not 100% certain about the findings. But how certain do we want to be? In geographical studies it is important to state the level of confidence before testing the hypothesis. The 95% level of confidence is usually used. Sometimes called the 0.05 probability level, it means that in the long run on five occasions out of one hundred we would reject the null hypothesis when we should not have done. In other words, we can be 95% certain that the findings are true and correct.

The decision to reject or accept the null hypothesis is usually made on the basis of comparing a calculated value with a value given in a set of special tables. Depending on the test used, the calculated value may have to be larger or smaller than the value in the table and we can reject the

TABLE 5.18

Number of days when midnight temperatures fall into the following 3 categories within and outside an urban area

°C	Within	Outside
5–8	27	196
8–10	93	84
10–12	245	85
Total	365	365

null hypothesis in favour of the alternative hypothesis. We can use a rule of thumb guide which suggests that our sample, for many situations, should contain between 30 and 50 values as a minimum.

Drawing conclusions. Remember that statistical analysis is only a tool in geography and is no substitute for good geographical reasoning. If the null hypothesis is not rejected, this may be either because there are other more important factors which we have not examined or that the sample was not large enough, or that there really is no difference between the samples. The test result will not help find which of these alternatives might be correct. A test on temperature data, such as that used above, is one example of an experiment set up on the basis of sound reasoning. However, as suggested in Chapter 1, incorrect conclusions can be drawn because the data are collected incorrectly or because we have not considered the problem properly. Many statistical textbooks show examples of strange relationships which may exist. For example, a statistically significant association may exist between the penguin population of Antarctica and the production of steel in Europe over the last 20 years. However, as sensible geographers, we would have had no reason for suspecting that such a relationship existed and would not have conducted the test in the first place! If hypotheses are stated clearly, then there is no reason why these so called *spurious relationships* should occur.

Tests of significance. The choice of statistical test depends on whether we are trying to find associations, differences or correlations between one, two or more sets of data. Table 5.19 gives some guidance as to the appropriate choice of test for given situations. The application of each different test is discussed below.

TABLE 5.19 Uses of inferential statistics

Levels of measurement	Tests of comparison		Tests of association		
	One sample Chi-squared test	Two or more sample Chi-squared test	Spearman's rank correlation analysis	Pearson's product – moment correlation analysis	Regression analysis
Nominal	√	√	×	×	×
Ordinal	√	√	√	×	√
Interval/ratio	√	√	√	√	√
Notes	Non-parametric test. Data grouped on nominal scale. Data must be in frequencies, not percentages. Compares an observed sample with a theoretical distribution.	Same requirements as one-sample test. Compares two or more sample distributions.	Non-parametric test. Interval and ratio data must be ranked. Examines the relationship between two variables. Student's t test is used to assess significance.	Parametric test and so assumes data are normally distributed. Only interval/ratio data. Otherwise same as Spearman's rank correlation analysis.	Used only when a significant linear correlation has been found. Informs about the strength of relationship between two variables. Allows predictions to be made about data.

Tests of comparison

The Chi-squared test (X²). This is a useful method for comparing data. Two types of comparison are commonly needed by geographers:

  i comparison of one set of data and a theoretical frequency distribution;

  ii comparisons of two or more sets of data, to see whether any differences exist.

The Chi-squared test can be used for both these situations, providing that the data are grouped in the form of *frequencies*, and the total *sample size is more than 20*. No assumption is made about data normality.

The one-sample test. This allows comparison of a set of data with a set of values we would expect for a given area or situation. It is a test of *goodness of fit* between an observed set of frequencies produced by a sample and a theoretical or expected set of frequencies. For example, it can be used to see whether a particular variable, such as the number of streams per km^2, is evenly distributed in a number of areas with different characteristics, such as geology. Table 5.20 shows the number of streams found in four areas of different geology. The one-sample X^2 test will enable us to see whether the numbers in each of the geological groups could have occurred by chance. Follow the steps given below:

1 State the null and alternative hypotheses.

Null hypothesis: There is no association between geology and number of streams.

Alternative hypothesis: There is a significant association between geology and stream numbers.

2 State the level of significance at which the null hypothesis will be accepted.

Null hypothesis will be tested at the 95% confidence level.

3 Calculate the number of streams we would expect to find in each area if geology had no influence. (If more than 20% of the expected values are less than 5 and if any of the expected values are less than 1, the data should be rearranged into a smaller number of groups.)

The number of streams we would expect in each area is calculated by dividing the total number of streams (130) by the number of groups of data (4 rock types). In each case, we expect 32.5 streams for each rock type. Tabulate the observed and expected data:

Rock type	Observed Frequency	Expected Frequency	$\frac{(O-E)^2}{E}$
Chalk	11	32·5	$(11-32·5)^2/32·5 = 14·2$
Sandstone	29	32·5	$(29-32·5)^2/32·5 = 0·4$
Slate	37	32·5	$(37-32·5)^2/32·5 = 0·6$
Quartzite	53	32·5	$(53-32·5)^2/32·5 = 12·9$
			$\sum \frac{(O-E)^2}{E}$ Total 28·1

4 Calculate the test statistic from the general formula given in Box 8.

5 Look up the calculated value in a table of significance. In order to do this the calculated value is compared against a set of tabulated values. The horizontal axis of the table sets out the confidence levels, whereas the vertical axis is defined by *degrees of freedom*. Degrees of freedom represent the size of the sample. For smaller samples, larger critical values of the test statistic will need to be reached before the null hypothesis can be rejected. In this case degrees of freedom are calculated as the number of rock types minus 1, i.e. 3. The tabulated X^2 value for 3 degrees of freedom at the 95% confidence level is 7.815.

6 Reject or accept the null hypothesis. Since the calculated value (28.1) *is greater* than the tabulated value (7.815) the *null hypothesis is rejected* in favour of the alternative hypothesis at the 95% confidence level.

TABLE 5.20

Rock type	No. of streams
Chalk	11
Sandstone	29
Slate	37
Quartzite	53
	Total 130

TABLE 5.21
Relative humidity data for two stations

Relative humidity %	Near sea	Away from sea
50–55	6	35
55–60	17	16
60–65	26	3
	$\Sigma 49$	$\Sigma 54$

TABLE 5.22
Contingency Table

	Column 1 (C_1)	Column 2 (C_2)	
Row 1 (r_1)	r_1c_1	r_1c_2	Σr_1
Row 2 (r_2)	r_2c_1	r_2c_2	Σr_2
Row 3 (r_3)	r_3c_1	r_3c_2	Σr_3
Totals	Σc_1	Σc_2	Grand Total $\Sigma r \Sigma c$

Two or more sample test. On some occasions we may need to know whether there is a difference between two or more samples. For example, we might wish to establish whether the relative humidity of the air is generally higher in coastal rather than inland locations. Like the one-sample test, data are arranged as frequencies such as that shown in Table 5.21. Note that one of the scales is nominal (i.e. near to or far away from the sea) and the other is interval. Like the one-sample test, we make no assumptions about data normality, but if more than 20% of the expected values fall below 5 and if any are less than 1, the data must be rearranged into a number of smaller groups. Follow the same steps as given for the one-sample test, but calculate the statistic by the method given in Box 9. The data are arranged into columns and rows, called a *contingency table* (Table 5.22). A contingency table comprises a number of cells (eg. r_1c_1, r_2c_1, etc.). In this example, there are *two* columns, *three* rows and *six* cells.

1 Calculate the expected frequency for each cell. This is achieved by using the following formula:

$$\text{Expected frequency for cell 1 } (r_1c_1) = \frac{\text{column total} \times \text{row total}}{\text{grand total}}$$

This can be shown in notation: $Er_1c_1 = \dfrac{\Sigma c_1 \times \Sigma r_1}{\text{grand total}}$

where: Er_1c_1 = expected frequency of a cell

Σc_1 = column total

Σr_1 = row total

In this case $Er_1c_1 = \dfrac{(49 \times 41)}{103} = 19.5$

The expected values are calculated for all row and column frequencies as shown in Table 5.22.

TABLE 5.23

	Near sea (**c.**)		Away from sea (**c.**)		
	Obs	Exp	Obs	Exp	Total
Row 1 (r^1)	6	19·5	35	21·5	41
Row 2 (r^2)	17	15·7	16	17·3	33
Row 3 (r^3)	26	13·8	3	15·2	29
	49		54		Grand Total 103

2 From Table 5.23, we calculate the X^2 value according to Box 9, but using the same procedure and assumptions shown for the one-sample test above. The following calculations are made:

$$\frac{(6 - 19 \cdot 5)^2}{19 \cdot 5} + \frac{(35 - 21 \cdot 5)^2}{21 \cdot 5} + \frac{(17 - 15 \cdot 7)^2}{15 \cdot 7} + \frac{(16 - 17 \cdot 3)^2}{17 \cdot 3} + \frac{(26 - 13 \cdot 8)^2}{13 \cdot 8} + \frac{(3 - 15 \cdot 2}{15 \cdot 2}$$

$$= 9 \cdot 3 + 8 \cdot 5 + 0 \cdot 1 + 0 \cdot 1 + 10 \cdot 9 + 9 \cdot 8 = 38 \cdot 6$$

$$X^2 = 38 \cdot 6$$

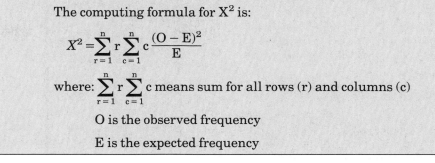

3 In this analysis, the degrees of freedom are the number of rows (r) minus 1 times the number of columns (c) minus 1, i.e.

df = (r − 1) × (c − 1)
df = (3 − 1) × (2 − 1) = 2

The calculated value of the test is 38·6.

The tabulated value at the 95% confidence level and at 2 degrees of freedom is 5·99.

Since the *calculated value* is greater than the *tabulated value*, we can *reject* the null hypothesis and accept the alternative hypothesis which states that there is an association between proximity to sea and relative humidity.

On the basis of the test statistic, the null hypothesis has been rejected. It is important that some attempt is made to explain why this occurs. Go back and look at the original table (Table 5.21) and it becomes clear that most values near the sea fall into the highest humidity class and that most of the values far from the sea fall into the low humidity class. Although a straightforward interpretation of this result can be given, remember that it is important to re-examine the table once the test statistic has been calculated because it will help with the interpretation of the results.

Tests of association

Correlation and regression. These are statistical techniques which show how well two variables relate to each other. The term variable means a factor or a measure , such as soil depth or population size or temperature. In some cases, an increase in the numerical value of one of the variables corresponds to an increase in the numerical value of a second variable. For example, we might expect that on a regional basis as altitude increases, annual rainfall also increases. In other situations, as one variable increases another decreases; for example, as we move further from the centre of a city, the rateable value of retail shops might decrease. These relationships are shown schematically in a graph called a *scattergram*, which plots a point at the intersection of the values along the x and y axis of the graph. The data of Tables 5.24 and 5.25 can be used to illustrate these features.

The data from these two tables are plotted in Figure 5.18. Figure 5.18 A shows a positive association, that is, both values increase together. Figure 5.18 B shows a negative relationship where an increase in distance from the city centre is matched by a decrease in the rateable value of retail shops.

Correlation is a statistical measure which allows us to gauge how closely one variable is related to another. Look at the 5 diagrams in Figure 5.19. These are five typical situations you might find with different sets of geographical data. Sometimes the relationship between two variables is quite close, such as in A and C. On other occasions, we get a range of values on the y axis for the same value of x, as in B and D. These latter graphs have a lower correlation. Graph E shows two variables which show no

TABLE 5.24
Annual rainfall values for seven stations as related to altitude

Altitude m	Annual Rainfall mm
50	350
78	425
103	498
137	531
189	629
249	700
362	980

TABLE 5.25
Distance from city centre as related to the rateable value of retail shops

Distance from centre km	Rateable value £
0	3500
0·2	3400
0·5	3200
1·0	3000
1·5	2600
2·2	2200
2·8	1900
2·9	1900
3·5	1500
5·0	610

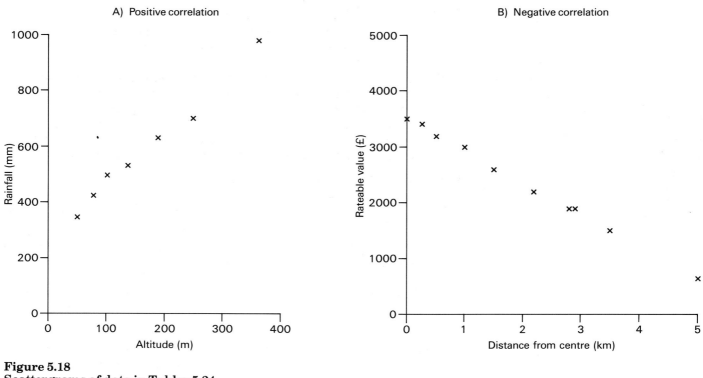

Figure 5.18
Scattergrams of data in Tables 5.24
and 5.25

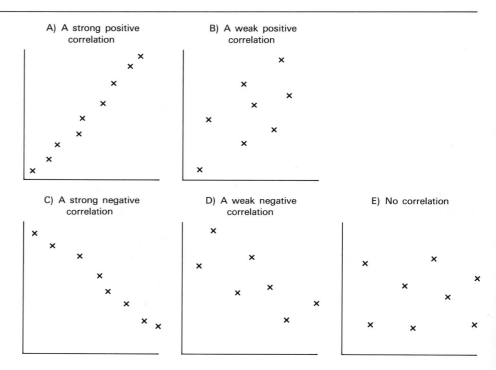

Figure 5.19
Strength of correlation

correlation. The *correlation coefficient* gives an exact measure of the degree of association between two variables. Two different correlation techniques can be used. If the data set does not have a normal frequency distribution, the *Spearman's rank correlation* is used. Where the data have a normal frequency distribution, *Pearson's product-moment correlation* should be used. In both cases, we can conduct a test of significance to see whether the relationship could have occurred by chance.

Spearman's rank correlation. This test calculates a value to show the degree of association between the two variables. The value ranges between 0 and +/− 1.0. Zero indicates no correlation and 1.0 a perfect correlation. The value is +/− depending on the direction of association.

Correlation	$+1.0 \longleftarrow$	$\longrightarrow 0.0 \longrightarrow$	$\longrightarrow -1.0$
	Perfect positive correlation	No correlation	Perfect negative correlation

Rank correlation can be performed on data measured on the ordinal, interval or ratio scales, although it needs reorganising, as the name implies, into a ranked format if measured at a level higher than the ordinal scale. For example, the data of Table 5.26 taken from a farm survey, enables us to relate average field size to altitude. Can we detect a correlation in the data? The following procedure is recommended:

1 State the null and alternative hypotheses.

Null hypothesis: There is no significant association between field size and altitude at the 95% significance level.

Alternative hypothesis: There is a significant association between field size and altitude at the 95% significance level.

2 The data are converted from the original ratio scale to an ordinal scale by ranking the numbers from highest (rank 1) to lowest (rank n) as shown in Table 5.26.

3 From the above table, the ranked data are substituted into the formula given in Box 10 to calculate the rank correlation coefficient.

TABLE 5.26
Field size and altitude

Altitude (m)	Rank	Field size (ha)	Rank
25	7	·8	1
120	5	·4	3
30	6	·7	2
190	2	·2	5
180	3	·25	4
179	4	·1	6
350	1	·05	7

BOX 10   Spearman's rank correlation coefficient

$$r_s = 1 - \left(\frac{6 \sum_{i=1}^{n} d^2}{(n^3 - n)} \right)$$

Where: r_s = Spearman's rank correlation coefficient

$\sum_{i=1}^{n}$ d = the difference in rank between the two data sets

n = number of paired observations

TABLE 5.27

Rank of altitude	Rank of field size	Difference	Difference squared
7	1	6	36
5	3	2	4
6	2	4	16
2	5	−3	9
3	4	−1	1
4	6	−2	4
1	7	−6	36
		Total	Σ106

4 From Table 5.26, it is possible to calculate the values required for the formula in Box 10 (Table 5.27).

By substituting values into the formula of Box 10, we get:

$$r_s = 1 - \left(\frac{6 \times 106}{(343 - 7)} \right)$$

$$= 1 - (636/336)$$

$$= 1 - 1 \cdot 893$$

$$= - 0 \cdot 893$$

5 Assess the significance of the correlation coefficient.
The correlation coefficient on its own is not a measure of significance, and it is not possible to judge whether the value is close enough to −1.0 to be *statistically significant*. Part of the problem relates to the effect of sample size. The larger the sample, the smaller the correlation coefficient needs to be in order to be statistically significant. A test of significance called the *Student's t test* is used. The computing formula is given in Box 11.

Significance of the correlation coefficient: Student's t test

$$t = r_s \times \sqrt{\left(\frac{n-2}{1-r_s^{\,2}}\right)}$$

where: r_s = Spearman's rank correlation coefficient
n = sample size
t = Student's t statistic

6 The calculated value of t is compared with the tabulated value in Student's t tables at the 95% significance level and at n−2 degrees of freedom.

By substituting the calculated value into Box 11, we get:

$$t = -\cdot89 \times \sqrt{\frac{5}{1 - \cdot797}}$$

$$= -\cdot89 \times \sqrt{\frac{5}{\cdot203}}$$

$$= -\cdot89 \times \sqrt{24\cdot63}$$

$$= -\cdot89 \times \ 4\cdot96$$

$$= -4\cdot42$$

At the 95% confidence level and at n−2 or 5 degrees of freedom, the tabulated value of t at the 95% confidence level is 2.571. Because the calculated value is *greater than* the tabulated value (ignore the negative sign here) we can reject the null hypothesis in favour of the alternative hypothesis which states that there is a significant negative relationship between altitude and field size.

On some occasions, rank correlation is less than satisfactory because we have a number of *tied ranks*. For example, if the numbers in the margin are ranked, three of them are the same. These three values are each assigned the same rank and the last value ranked 8 out of the 8 numbers. If one of the following conditions are found to exist a correction must be applied. These conditions are:

i when three or more observations are of the same rank;
ii when the number of tied pairs is more than 25% of the total sample size.

Since this situation occurs rarely, the special adjustment is not given here. Reference should be made to more advanced statistical texts for this correction.

Pearson's product-moment correlation coefficient. The most important requirement for this test is that the data have a normal frequency distribution. If the data are not normally distributed or if you are in doubt, then rank correlation should be used. Only interval and ratio scale data can be used. This correlation coefficient also has a range of +1 to −1 and is interpreted in the same way as r_s. The procedure for setting up hypotheses should follow that set out for rank correlation, but note that the computing formula of Box 12 is very different.

Number	Rank
17	1
12	4
11	5
14	3
11	5
16	2
11	5
9	8

132

$$r = \frac{n \times \Sigma XY - (\Sigma X \times \Sigma Y)}{\sqrt{[n \times \Sigma X^2 - (\Sigma X)^2] \times [n \times \Sigma Y^2 - (\Sigma Y)^2]}}$$

Where: r = Pearson's correlation coefficient
Σ = sum from i=1 to n (i=1 to n has been omitted to simplify the formula here)
X = each observation of the first variable (such as altitude)
Y = each observation of the second variable (such as field size)

The data from Table 5.26 can again be used to illustrate the calculation. It is necessary to sum the following values:

1 the values of X
2 the values of Y
3 the values of X^2 X means variable X (altitude)
4 the values of Y^2 Y means variable Y (field size)
5 the values of $X \times Y$

TABLE 5.28

	X	Y	X^2	Y^2	$X \times Y$
	25	·80	625	·640	20
	120	·40	14400	·160	48
	30	·70	900	·490	21
	190	·20	36100	·040	38
	180	·25	32400	·063	45
	179	·10	32041	·010	17·9
	350	·05	122500	·003	17·5
Totals	1074	2·50	238966	1·406	207·4

Sample size (n) = 7

By substituting the values from Table 5.28 into the formula of Box 12, we get:

$$r = \frac{(7 \times 207\cdot4) - (1074 \times 2\cdot5)}{\sqrt{[(7 \times 238966) - 1153476] \times [(7 \times 1\cdot406) - 6\cdot25)]}}$$

$$= \frac{1451\cdot8 - 2685}{\sqrt{519286 \times 3\cdot592}}$$

$$= \frac{-1233\cdot2}{1365\cdot75}$$

$$= -\cdot903$$

The significance test given in Box 11 can be used to test the null and alternative hypotheses of this relationship in the same way as for rank correlation explained above. The result becomes:

$$t = -\cdot903 \times \sqrt{\frac{5}{1 - (-\cdot903)^2}}$$

$$= -\cdot093 \times \sqrt{\frac{5}{\cdot1846}}$$

$$= -0\cdot903 \times 5\cdot204$$

$$= -4\cdot699$$

As in the above rank correlation example, the degrees of freedom are n−2 (5). The tabulated value in the Student's t table is 2.571. Since 4.699 is

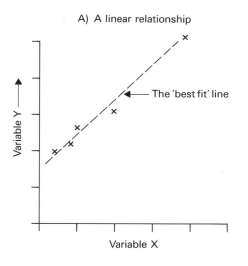

A) A linear relationship

The 'best fit' line

Variable Y

Variable X

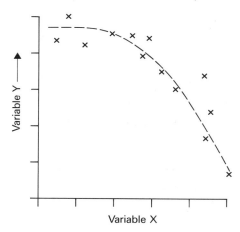

B) A non-linear relationship

Variable Y

Variable X

Figure 5.20
Linear and non-linear regression

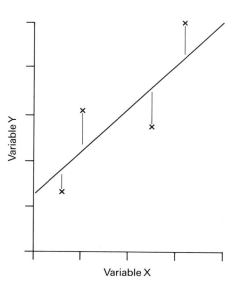

Variable Y

Variable X

Figure 5.21
Definition of least square distance

greater than 2.571, we can *reject* the null hypothesis in favour of the alternative hypothesis which states that there is a significant relationship between field size and altitude.

Interpretation of the result requires not only a statement that the relationship is significant, but some attempt to make a more detailed assessment. For example, the value of the correlation coefficient is negative. This means that as altitude increases, field size decreases. There may be several reasons for this which may relate to economic and physical factors such as:

i   the need to protect crops by hedgerows and small fields at high altitudes;
ii the difficulty of using large machinery on steep upland slopes;
iii little investment in agriculture in low productivity upland areas.

Remember that the interpretation is still a matter of judgement and skilled argument. The statistical analysis can only help by providing some weight to the significance of a particular relationship or set of relationships you have chosen to examine.

Regression analysis. Correlation analysis tells us something about the strength of a relationship between two variables. Regression analysis allows us to predict the value of one variable from another variable and identifies values which deviate from the average relationship (*residuals*). It is only worth undertaking a regression analysis if:

i the two variables have a significant correlation;
ii the relationship between the two variables is linear.

Look at Figure 5.20. In example A, the two variables when plotted produce an almost perfect straight line on the graph. The dotted line is the *best fit* regression line which is linear and is positioned in such a way that it best represents all of the points on the graph (we will look at its precise definition later). In graph B the relationship between the two variables cannot be represented by a straight line, and the curved dotted line best describes the average relationship between the two variables. This form of relationship should not be tackled by linear regression analysis.

The regression line is fitted to the data in one of two ways. Either the position of the line is estimated by eye or is positioned by calculation. In the second case, it is done in such a way so that the average distance between the line and the plotted points is at a minimum. Figure 5.21 shows a plotted regression line and 4 points surrounding the regression line. The distance between each point and the predicted line is measured parallel to the y axis. If the squared distances between all points and the regression line are summed the value is smaller than for any other line plotted through the same points, i.e.

$$\text{min} = (Yo{-}Yp)^2$$

where: Yo = the observed value of Y
       Yp = predicted value of Y from the regression line

The regression line is usually expressed as a simple formula given in Box 13.

> **BOX 13   The regression equation**
>
> $Y = a + bX$
>
> Y = the predicted values of the Y variable
> X = the observed values of the X variable
> a and b are constants

We can see from Figure 5.22 that 'a' is a point on the Y axis where the regression line crosses this axis at the point where x = 0. From the same diagram, we can see that 'b' is the slope or gradient of the regression line. Like correlation, regressions can be positive or negative, strong or weak.

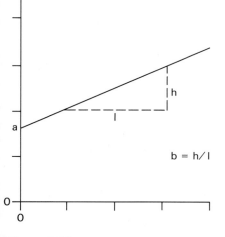

**Figure 5.22
Definition of regression equation**

**Figure 5.23
Strength of regression
relationships**

Examples of a variety of different strengths and directions of regression relationships are given in Figure 5.23.

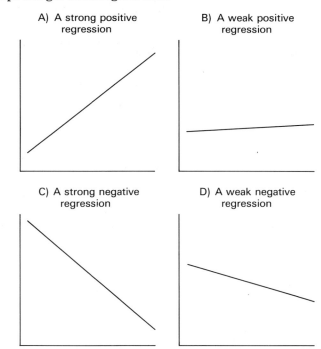

The values of 'a' and 'b' in the regression equation of Box 13 can be calculated by using the formula given in Box 14.

BOX 14   Calculation of the regression constants 'a' and 'b'

$$b = \frac{n \times \Sigma XY - [(\Sigma X) \times (\Sigma Y)]}{(n \times \Sigma X^2) - (\Sigma X)^2}$$

$$a = \frac{\Sigma Y - (b \times \Sigma X)}{n}$$

Where all values are defined in Box 12 and Table 5.26.

The data set out in Table 5.26 for correlation analysis can be used to calculate the value of 'a' and 'b' according to the formula given in Box 14. For b:

$$b = \frac{(7 \times 207 \cdot 4) - (1074 \times 2 \cdot 5)}{(7 \times 238,966) - (1074 \times 1074)}$$

$$b = \frac{1451 \cdot 8 - 2685}{1672762 - 1153476}$$

$$b = \frac{-1233 \cdot 2}{519286}$$

$$b = -\cdot 0024$$

For a:

$$a = \frac{2 \cdot 5 - (-\cdot 0024 \times 1074)}{7}$$

$$a = \frac{2 \cdot 5 - (-2 \cdot 5776)}{7}$$

$$a = \frac{5 \cdot 0776}{7}$$

$$a = \cdot 725$$

135

The next step is to plot the regression line on a scattergram showing the relationship between altitude and field size. Figure 5.24 shows the plotted data. The position of 'a' is marked on the y axis at the point where X = 0. For any value of X, calculate what Y should be by substituting the value of X into the formula of Box 13.

For a value of X = 200, it becomes:

$$Y = 0.725 + (-.0024 \times 200)$$
$$Y = 0.725 - .48$$
$$Y = 0.245$$

For an X value of 200, plot the value of Y = 0.245 on the graph (see Figure 5.24). Draw a straight line through this point to 'a'. This is the position of the regression line. You will find that the regression line goes through the point representing the mean of both altitude and field size (Figure 5.24).

After fitting the regression line it is possible to predict any value of Y for a given value of X, either by reading values from the graph or by using the formula of Box 13. We can also see which points do not fit the regression line very well. This is called *analysing the residuals from the regression*. For example, a field size of 0.1 ha at an altitude of 179m on Figure 5.24 is much smaller than we would have expected. It is possible to re-examine this site to find out why this might have happened. Analysis of these residual points can often provide a far more interesting exercise than discussing an expected relationship.

A note of caution. Correlation and regression are very powerful statistical techniques, but no amount of statistical analysis can be seen as a substitute for sound geographical reasoning. A high correlation and a strong regression are often assumed to mean that variable X *causes* variable Y. This is not true. Correlation and regression establish the likelihood of a relationship occurring by chance but still do not prove cause and effect. The exact cause may be through some other variable which relates to the ones we have measured but which we have not thought about in our analysis. In the above example, our conclusion is that there is a strong relationship between altitude and field size. The reason for this trend may be very complex, because at higher altitudes there may be steeper slopes which are difficult to plough, or the sites are so exposed that the farmer has small fields with high hedges to protect his crops. Any piece of work should not simply present the result of a statistical test, but should consider the implications of the results of such a test, paying special attention to residual values.

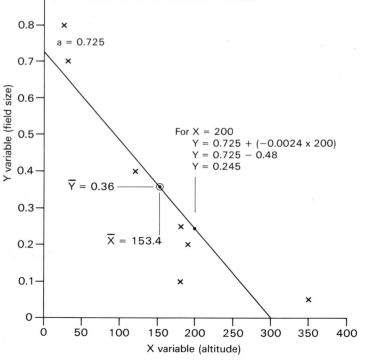

Figure 5.24

136

EXERCISES: INFERENTIAL STATISTICS

EXERCISE 1:   **CHI-SQUARED ANALYSIS, ONE-SAMPLE TEST**

Table 5.29 relates to a vegetation survey which has identified the number of different species present in a sample of woodland, parkland, grassland and urban scrubland sites.

You have to establish whether different sites have more or less species than would be expected for an even distribution of habitats. Use the one-sample Chi-squared test and:

Q **1** State the null and alternative hypotheses.

2 Calculate the expected frequencies.

3 Calculate the Chi-squared statistic.

4 Compare the calculated and tabulated value at the appropriate degrees of freedom.

5 Decide whether you can reject the null hypothesis.

6 Explain the result you have obtained.

TABLE 5.29

Site Type	No of Species
Woodland	53
Grassland	23
Parkland	9
Urban scrubland	15
Total	100

EXERCISE 2:   **CHI-SQUARED ANALYSIS, TWO-SAMPLE TEST**

The data in Table 5.30 represent measurements of pebble size at two points (A and B) on a beach which is strongly influenced by longshore drift (Figure 5.25). The pebble sizes are classified into three groups.

Use the two-sample Chi-squared method to calculate the test statistic.

Q **1** State the null and alternative hypotheses.

2 Calculate the expected frequencies.

3 Calculate the Chi-squared statistic.

4 Compare the calculated and tabulated value at the appropriate degrees of freedom.

5 Decide whether or not to reject the null hypothesis.

6 Explain the result you have obtained.

TABLE 5.30

Size (mm)	Site A	B	Total
0–3	3	11	14
3–6	7	6	13
6–9	10	3	13
Totals	20	20	40

Figure 5.25

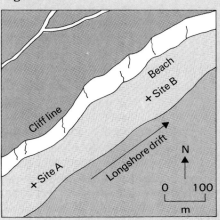

SPEARMAN'S RANK CORRELATION

The data in Table 5.31 are based on a sample of European statistics relating infant mortality rate per 1000 head of population to an index of wealth (GNP per capita). The raw data are not available, and countries have been ranked from lowest to highest on the infant mortality and wealth scales.

TABLE 5.31

Country	Infant mortality rank	Wealth rank	Country	Infant morality rank	Wealth rank
Albania	1	15	Bulgaria	9	11
Yugoslavia	2	13	Austria	10	6
Portugal	3	14	W Germany	11	3
Poland	4	9	United Kingdom	12	4
Hungary	5	8	Denmark	13	2
Spain	6	12	Finland	14	5
Italy	7	7	Sweden	15	1
Greece	8	10			

Use Spearman's rank correlation method.

Q 1 State the null and alternative hypotheses.

2 Calculate the correlation coefficient.

3 Calculate the t statistic.

4 Compare the calculated value of t with the tabulated value at the appropriate degrees of freedom.

5 Decide whether to accept or reject the null hypothesis.

6 Explain your findings.

PEARSON'S PRODUCT-MOMENT CORRELATION

The data in Table 5.32 are derived from a shopping survey which looked at the price of a basket of groceries bought in different shopping centres in a large urban area. Data were also compiled on the distance of each shopping centre from the middle of town.

Use the product-moment correlation method.

Q 1 State the null and alternative hypotheses.

2 Calculate the correlation coefficient.

3 Calculate the value of Student's t.

4 Compare the calculated value with the tabulated value at the appropriate degrees of freedom.

5 Decide whether to accept or reject the null hypothesis.

If the relationship is significant:

6 Plot a scattergram of distance (x axis) against price of basket of goods (y axis).

7 Calculate the regression equation.

8 Fit the regression line.

9 Explain the results of the correlation and regression analysis.

TABLE 5.32

Price for basket of goods £	Distance from city centre km
13.70	0.00
13.80	0.01
14.25	0.50
14.70	1.60
14.90	3.00
15.11	3.50
15.37	4.80
15.84	6.00

INDEX